IMPROVING DATA TO ANALYZE
Food and Nutrition Policies

Panel on Enhancing the Data Infrastructure in Support of Food and Nutrition Programs, Research, and Decision Making

Committee on National Statistics

Division of Behavioral and Social Sciences and Education

NATIONAL RESEARCH COUNCIL
OF THE NATIONAL ACADEMIES

THE NATIONAL ACADEMIES PRESS
Washington, DC
www.nap.edu

THE NATIONAL ACADEMIES PRESS   500 Fifth Street, N.W.   Washington, DC 20001

NOTICE: The project that is the subject of this report was approved by the Governing Board of the National Research Council, whose members are drawn from the councils of the National Academy of Sciences, the National Academy of Engineering, and the Institute of Medicine. The members of the committee responsible for the report were chosen for their special competences and with regard for appropriate balance.

This study was supported by Contract/Grant No. 43-3AEM-3-80119 between the National Academy of Sciences and the U.S. Department of Agriculture. Support of the work of the Committee on National Statistics is provided by a consortium of federal agencies through a grant from the National Science Foundation (Number SBR-0112521). Any opinions, findings, conclusions, or recommendations expressed in this publication are those of the author(s) and do not necessarily reflect the views of the organizations or agencies that provided support for the project.

International Standard Book Number 0-309-10005-4

Additional copies of this report are available from the National Academies Press, 500 Fifth Street, N.W., Lockbox 285, Washington, DC 20055; (800) 624-6242 or (202) 334-3313 (in the Washington metropolitan area); Internet, http://www.nap.edu

Printed in the United States of America
Copyright 2005 by the National Academy of Sciences. All rights reserved.

Suggested citation: National Research Council. (2005). *Improving Data to Analyze Food and Nutrition Policies*. Panel on Enhancing the Data Infrastructure in Support of Food and Nutrition Programs, Research, and Decision Making. Committee on National Statistics, Division of Behavioral and Social Sciences and Education. Washington, DC: The National Academies Press.

THE NATIONAL ACADEMIES
Advisers to the Nation on Science, Engineering, and Medicine

The **National Academy of Sciences** is a private, nonprofit, self-perpetuating society of distinguished scholars engaged in scientific and engineering research, dedicated to the furtherance of science and technology and to their use for the general welfare. Upon the authority of the charter granted to it by the Congress in 1863, the Academy has a mandate that requires it to advise the federal government on scientific and technical matters. Dr. Ralph J. Cicerone is president of the National Academy of Sciences.

The **National Academy of Engineering** was established in 1964, under the charter of the National Academy of Sciences, as a parallel organization of outstanding engineers. It is autonomous in its administration and in the selection of its members, sharing with the National Academy of Sciences the responsibility for advising the federal government. The National Academy of Engineering also sponsors engineering programs aimed at meeting national needs, encourages education and research, and recognizes the superior achievements of engineers. Dr. Wm. A. Wulf is president of the National Academy of Engineering.

The **Institute of Medicine** was established in 1970 by the National Academy of Sciences to secure the services of eminent members of appropriate professions in the examination of policy matters pertaining to the health of the public. The Institute acts under the responsibility given to the National Academy of Sciences by its congressional charter to be an adviser to the federal government and, upon its own initiative, to identify issues of medical care, research, and education. Dr. Harvey V. Fineberg is president of the Institute of Medicine.

The **National Research Council** was organized by the National Academy of Sciences in 1916 to associate the broad community of science and technology with the Academy's purposes of furthering knowledge and advising the federal government. Functioning in accordance with general policies determined by the Academy, the Council has become the principal operating agency of both the National Academy of Sciences and the National Academy of Engineering in providing services to the government, the public, and the scientific and engineering communities. The Council is administered jointly by both Academies and the Institute of Medicine. Dr. Ralph J. Cicerone and Dr. Wm. A. Wulf are chair and vice chair, respectively, of the National Research Council.

www.national-academies.org

PANEL ON ENHANCING THE DATA INFRASTRUCTURE IN SUPPORT OF FOOD AND NUTRITION PROGRAMS, RESEARCH, AND DECISION MAKING

JOHN KARL SCHOLZ *(Chair)*, Department of Economics and Institute for Research on Poverty, University of Wisconsin–Madison
F. JAY BREIDT, Department of Statistics, Colorado State University
RONETTE BRIEFEL, Mathematica Policy Research, Inc., Princeton, New Jersey
WILLIAM F. EDDY, Department of Statistics, Carnegie Mellon University
ANDREW GELMAN, Department of Statistics and Department of Political Science, Columbia University
ALAN R. KRISTAL, Fred Hutchinson Cancer Research Center, University of Washington
BARRY M. POPKIN, Division of Nutrition Epidemiology, School of Public Health, University of North Carolina
LAURIAN J. UNNEVEHR, Department of Agricultural and Consumer Economics, University of Illinois at Urbana-Champaign
WALTER WILLETT, Department of Nutrition, School of Public Health, Harvard University, and Harvard Medical School

EARL S. POLLACK, *Study Director*
MICHELE VER PLOEG, *Study Director (until October 2004)*
JAMIE CASEY, *Research Associate*
TANYA M. LEE, *Program Assistant (until October 2004)*
MICHAEL SIRI, *Senior Program Assistant*

COMMITTEE ON NATIONAL STATISTICS
2004-2005

WILLIAM F. EDDY (*Chair*), Department of Statistics, Carnegie Mellon University
KATHARINE ABRAHAM, Joint Program in Survey Methodology, University of Maryland
ROBERT BELL, AT&T Research Laboratories, Florham Park, New Jersey
LAWRENCE D. BROWN, Department of Statistics, The Wharton School, University of Pennsylvania
ROBERT M. GROVES, Survey Research Center, University of Michigan, and Joint Program in Survey Methodology
JOHN HALTIWANGER, Department of Economics, University of Maryland
PAUL W. HOLLAND, Educational Testing Service, Princeton, New Jersey
JOEL L. HOROWITZ, Department of Economics, Northwestern University
DOUGLAS MASSEY, Department of Sociology, Princeton University
VIJAY NAIR, Department of Statistics and Department of Industrial and Operations Engineering, University of Michigan
DARYL PREGIBON, Google, Inc., New York, New York
KENNETH PREWITT, School of International and Public Affairs, Columbia University
LOUISE RYAN, Department of Biostatistics, Harvard University
NORA CATE SCHAEFFER, Department of Sociology, University of Wisconsin–Madison

CONSTANCE F. CITRO, *Director*

Acknowledgments

On behalf of the panel, I would like to thank all of the individuals involved in the production of this report. First, I thank my fellow panel members for giving their time and expertise so generously toward the completion of this report. Each of their contributions to the discussion at the workshop and in the drafting of this report is greatly appreciated.

A major activity of this panel was a workshop, held in May 2004. I thank everyone who presented information and shared their ideas for improvements in the food consumption and expenditures data infrastructure. Their willingness to share these ideas in a somewhat informal manner during the workshop was very much appreciated. I appreciate the help of several individuals from the federal government, private firms, and universities who were consulted by staff of the Committee on National Statistics about workshop topics and potential workshop participants: Harry Balzer of the NPD Group; Mary Brandt of the Food and Drug Administration; Gary Crisafulli of ACNielsen; Helen Jensen of Iowa State University; Clifford Johnson of the National Center for Health Statistics, who also provided critical information on current details and future plans for the National Health and Nutrition Examination Survey; Alanna Moshfegh of the Agricultural Research Service of the U.S. Department of Agriculture; and Catherine Woteki of the University of Iowa. The summary of the workshop is Appendix A.

I also thank our sponsor, the Economic Research Service (ERS) of the U.S. Department of Agriculture. ERS staff Elizabeth Kuhn, Nicole

Ballenger (who is now with the University of Wyoming), James Blaylock, Mark Denbaly, and Biing-Hwan Lin aided the development of the workshop and provided information to the panel.

The panel is grateful for the excellent work of the staff of the Committee on National Statistics (CNSTAT) and the National Research Council for developing and organizing the workshop and this report. Michele Ver Ploeg, study director for the panel until October 2004, was chiefly responsible for organizing the workshop and drafting the panel's report. Earl Pollack, study director for the panel since October 2004, completed the report and shepherded it through the review process with the assistance of Constance Citro, CNSTAT director. Throughout, the panel benefited immensely from Connie's advice and collaboration. We also thank Jamie Casey, research associate, for drafting the workshop summary and for providing research support, and Tanya Lee, program assistant, for handling all administrative matters for the workshop. Jerusha Nelson Peterman, science and technology policy intern with CNSTAT, gathered background information on many of the datasets discussed in the report and contacted staff at the agencies that produce these data to clarify details. The panel is also grateful to Eugenia Grohman of the reports office of the Division of Behavioral and Social Sciences and Education for her professional editing of the report.

This report has been reviewed in draft form by individuals chosen for their diverse perspectives and technical expertise, in accordance with procedures approved by the Report Review Committee of the National Research Council. The purpose of this independent review is to provide candid and critical comments that will assist the institution in making its published report as sound as possible and to ensure that the report meets institutional standards for objectivity, evidence, and responsiveness to the study charge. The review comments and draft manuscript remain confidential to protect the integrity of the deliberative process.

We thank the following individuals for their review of this report: Jay Bhattacharya, Stanford Medical School, Stanford University; Gladys Block, Public Health Nutrition Program, School of Public Health, University of California, Berkeley; Daniel S. Gaylin, Department of Health Survey, Program, and Policy Research, National Opinion Research Center, Washington, DC; Chris Logan, Education and Family Support, Abt Associates Inc., Cambridge, MA; Valerie Tarasuk, Department of Nutritional Sciences, University of Toronto; Parke E. Wilde, Friedman School of Nutrition

Science and Policy, Tufts University; and James P. Ziliak, Center for Poverty Research and Department of Economics, University of Kentucky.

Although the reviewers listed above have provided many constructive comments and suggestions, they were not asked to endorse the conclusions or recommendations nor did they see the final draft of the report before its release. The review of this report was overseen by Robert Moffitt, Department of Economics, Johns Hopkins University. Appointed by the National Research Council, he was responsible for making certain that an independent examination of this report was carried out in accordance with institutional procedures and that all review comments were carefully considered. Responsibility for the final content of this report rests entirely with the authoring panel and the institution.

> John Karl Scholz, *Chair*
> Panel on Enhancing the Data Infrastructure
> in Support of Food and Nutrition Programs,
> Research, and Decision Making

Contents

Executive Summary — 1

1 Introduction — 7
   Panel Charge and Context, 10
   Food Consumption Decisions, 18
   Food Safety Issues, 21
      Food-Borne Pathogens, 22
      Pesticide Residues, 24
      Evaluating Consumer Education, 25

2 Federal Datasets on Food and Nutrition — 26
   National Health and Nutrition Examination Survey, 26
   Continuing Survey of Food Intakes by Individuals, 38
   Integrated NHANES and CSFII, 39
      Limitations, 39
         Exclusion of Some CSFII Content, 39
         Limited Economic Data, 40
         Design Limitations, 40
      Potential for Improved Data, 41
         ERS Supplement Initiative, 42
         Special Supplements, 43
         Links to Food Assistance Program Records, 44
         Links to Geographic Databases on Food Outlets, 46

xi

        Links to Neighborhood Characteristics, 46
        Links to Price Information, 47
        Confidentiality and Data Access, 49
    Consumer Expenditure Survey, 50
        Uses and Limitations, 51
        Possible Improvements, 52

3   Proprietary Data Sources   53
    Scanner Data, 53
        Uses, 56
        Limitations, 57
        Future Potential, 59
    Household Food Consumption Surveys, 60

4   Other Federal Data Sources   63
    Current Population Survey, 64
    American Time Use Survey, 66
    Panel Surveys, 68
        Early Childhood Longitudinal Study, 68
        Health and Retirement Study, 69
        Panel Study of Income Dynamics, 71
    Quick-Turnaround Surveys, 72
        Behavioral Risk Factor Surveillance Survey, 72
        State and Local Area Integrated Telephone Survey, 74
    Expanded Food and Nutrition Education Program, 74

5   Recommendations   76
    Interagency Working Group on Food and Nutrition Data, 77
    Research and Development, 79
    Enhancing Food and Nutrition-Related Data in NHANES, 80
    Data Linkages, 81
    Use of Scanner Data, 82
    Use of Other Datasets, 83
    Conclusion, 84

References   85

Appendixes

A	Enhancing the Data Infrastructure in Support of Food and Nutrition Programs, Research, and Decision Making: Summary of a Workshop	91
B	Workshop Agenda	119
C	Workshop Participants	123
D	Biographical Sketches of Panel Members and Staff	128

Executive Summary

Several changes in the United States over the past two decades have implications for diet, nutrition, and food safety, including patterns of food consumption that have produced an increase in overweight and obese Americans and threats to food safety from pathogens and bioterrorism. The changes raise a number of critical policy and research questions: How do differences in food prices and availability or in households' time resources for shopping and food preparation affect what people consume and where they eat? How do factors outside of the household, such as the availability of stores and restaurants, food preparation technology, and food marketing and labeling policies, affect what people are consuming? What effects have food assistance programs had on the nutritional quality of diets and the health of those served by the programs? Where do people buy and consume food and how does food preparation affect food safety?

To address these and related questions, the Economic Research Service (ERS) of the U.S. Department of Agriculture (USDA) asked the Committee on National Statistics to convene a panel of experts to provide advice for improving the data infrastructure on food consumption and nutrition. The panel was charged to review data needs to support research and decision making for food and nutrition policies and programs in USDA and to assess the adequacy of the current data infrastructure and recommend enhancements to improve it. The panel was asked to consider improvements to current systems, not to propose major new systems. The panel's recommendations are based largely on the discussions at a workshop, which it sponsored in May 2004, to hear from USDA and other federal agencies with food and nutrition-related policy responsibilities and from statistical agencies and private firms that collect data on food consumption and expenditures.

FINDINGS: DATA SOURCES

No single data source currently provides or could provide all of the needed information. A number of data sources provide some of the information, but each has some weaknesses in addressing policy-related questions.

Relevant datasets fall under three categories:

1. federal datasets that are primary sources of data on food consumption, food expenditures, and dietary attitudes and knowledge:

- the National Health and Nutrition Examination Survey (NHANES), conducted by the National Center for Health Statistics on a continuing basis since 1999;
- the Continuing Survey of Food Intakes by Individuals (CSFII), last conducted by USDA in 1994-1996 (and again in 1998 for children under age 10), which was discontinued and then integrated into NHANES, beginning in 2002;
- the Diet and Health Knowledge Survey (DHKS), a past supplement to CSFII that was not part of the integrated NHANES-CSFII, but is a source of questions for a new supplement to NHANES under development by ERS, the Flexible Consumer Behavior Survey Module; and
- the Consumer Expenditure Survey (CE), conducted on a continuing basis since 1980 by the Bureau of Labor Statistics.

2. proprietary data collected by private market research firms to analyze food and related markets:

- retail and household scanner data, which include quantities sold and prices from bar codes on products purchased at retail outlets, for which the major producers are ACNielsen and Information Resources, Inc. (IRI);
- the National Eating Trends Survey, a small survey conducted by the NPD Group that obtains 14-day diaries of food intake; and

- the Consumer Report on Eating Share Trends (CREST), an online survey conducted by the NPD Group of people's previous-day purchases of prepared foods.

3. other federal datasets that could provide useful information:

- the monthly Current Population Survey (CPS), which includes a December module on food expenditures, food assistance program participation, and food insecurity;
- the American Time Use Survey (ATUS), conducted by the Bureau of Labor Statistics, a new survey that plans to add a food and eating module (in cooperation with ERS);
- longitudinal (panel) surveys that provide repeated measurements on the same individuals, permitting analysis of changing behavior over time, such as the Early Childhood Longitudinal Study, the Health and Retirement Study of people over age 50, and the Panel Study of Income Dynamics, the longest running nationwide survey of families' economic and demographic circumstances;
- surveys to which special modules to capture emerging trends can be added relatively easily, which include the Behavioral Risk Factor Surveillance System (BRFSS) and the State and Local Area Integrated Telephone Survey (SLAITS), both of which also provide state-level detail; and
- the Expanded Food and Nutrition Education Program (EFNEP), which helps low-income people provide nutritionally adequate meals for their families and collects information on their diet quality and food practices that may have research potential.

There are also several datasets that provide information relevant to food safety and awareness. The FoodNet system monitors outbreaks of food-related illness in ten sites. Periodically, the Food and Drug Administration asks people by telephone about food handling, food allergies, and consumption of potentially unsafe food in the Food Safety Survey and more broadly about awareness of relationships between diet and risks for chronic disease and health-related knowledge and attitudes in the Health and Diet Survey.

RECOMMENDATIONS

The multiplicity of sources of data related to food consumption, diet, and nutrition provide a range of information that is useful to policy makers and researchers at USDA and other agencies with food-related responsibilities, which include agencies in the Department of Health and Human Services and the Environmental Protection Agency. Yet that multiplicity also results in overlaps that are not efficient and in gaps that limit the information that policy makers and researchers have to address current and emerging issues in food and nutrition.

Responding to our charge to assess and consider improvements to the current data systems, we offer six recommendations to improve knowledge about the nation's changing patterns of food consumption. These broad recommendations are the basis for proposals for data enhancements throughout the report.

Recommendation 1: An interagency working group, led by the Office of Management and Budget, or co-led by an agency of the Department of Agriculture and the Department of Health and Human Services, should be established and take responsibility for the systematic development and use of diet and food consumption data to address policy and research questions of the federal government.

Recommendation 2: The proposed interagency working group should assign clear responsibilities to lead agencies for sustained programs of research and development on data in key areas to provide a sound base of scientific evidence for the group's work to improve the available information on diet and food consumption.

Recommendation 3: The proposed interagency working group on diet and food consumption data should consider priorities and methods for obtaining additional food and nutrition-related information in the National Health and Nutrition Examination Survey. The development of the NHANES Flexible Consumer Behavior Survey Module, which will include questions on food expenditures, diet and health knowledge, and other food and nutrition-related topics, should proceed, and research should be conducted on ways to obtain price information for inclusion in NHANES.

Recommendation 4: The proposed interagency working group on diet and food consumption data should consider low-cost ways to enhance the analytic uses of NHANES and other surveys by linkages with food assistance program records and with sources of socioeconomic and food shopping characteristics for the areas in which survey respondents live. A priority should be to work out effective ways to provide access to linked datasets through restricted access mechanisms, such as monitored remote on-line access.

Recommendation 5: The Economic Research Service of the U.S. Department of Agriculture should continue to explore the use of data on food purchases, prices, and consumption from proprietary retail scanner systems, household scanner panels, and household consumption surveys. This work should include a program to examine the quality of the data, consideration of ways to reduce the costs of access, and the determination of priority applications for the information.

Recommendation 6: The proposed interagency working group on diet and food consumption data should consider ways to enhance the usefulness of other federal datasets for food and nutrition-related policy analysis and research. Such datasets include the Current Population Survey, the American Time Use Survey, panel surveys that follow families, children, and the elderly over time, and surveys that are designed to include modules to track emerging trends.

1

Introduction

Changes in the U.S. public's food consumption and diet-related attitudes and information, together with advances in medical knowledge of dietary effects on health, have heightened awareness of the importance of understanding what people eat and why they eat it. At the same time, there are new challenges for diet, health, and food safety. Biotechnological innovations in food production have raised concerns among some consumers about the safety of some foods. Pathogens, such as *E. coli* O157:H7 and bovine spongiform encephalopathy (BSE) or mad cow disease, and the threat of terrorism have amplified concerns about food safety. And while nutrient deficiencies in the population remain, among the most pressing dietary problems today are overconsumption of trans and saturated fat, sodium, refined carbohydrates, and total calories, and underconsumption of fruits, vegetables, and whole grains (see www.ers.usda.gov/Briefing/DietAndHealth [June 2005]).

These developments raise important and intriguing policy and research questions. What has caused the increase in overweight and obese Americans? Are people eating more, eating the wrong foods, exercising less, or some combination of these? How do changes in food markets—food prices and availability—affect what people consume? How do other factors, such as income, time resources, and consumers' preferences and knowledge, affect food consumption decisions, and how have they changed over time? How do factors outside of homes, such as the availability of stores and restaurants, food and food preparation technology, food marketing and labeling policies, and incorporation of advances in dietary knowledge into health care delivery, affect what people are consuming and the consequences for their health and safety? Given that the prevalence of obesity is greater among low-income than other households, what effects have food assistance and educational programs had on the nutritional quality of diets of those served by the programs, and are these programs effective in improving diets and health? How do food consumption patterns affect food markets—for example, how do different weight-loss programs affect the purchasing and consumption of different foods? Where do people buy and consume food, and how does food preparation affect food safety? How does consumption of specific foods change after a food safety outbreak?

Many different kinds of information are needed to address these questions and to formulate or adjust policy: information on food expenditures, food consumption, food prices, where food is purchased and consumed, food preparation, diet and health knowledge, and possible sources of contamination. For example, to understand if foods with relatively high energy content and low nutrient content are being consumed rather than healthier foods because they are relatively cheaper or more readily available, data on food consumption, prices, and availability are needed. For some purposes, longitudinal data on the behavior of the same households over time are needed. For other purposes, data are needed on a very timely basis in order to make decisions based on current or recent market conditions.

While there are rich sources of data on food consumption and related issues, gaps exist, and no single source contains all of the information needed to answer these questions. For example, there are good data from the Consumer Expenditure Survey (CE) on food purchases by households. However, these data do not have information on who in a household consumes how much of the food; they do not contain detailed information on food consumed away from the home; and they do not include information on prices paid for specific quantities of particular foods. The National Health and Nutrition Examination Survey (NHANES) collects critical information on food consumption and health and nutritional status that has many uses for policy making and research. However, it does not now collect data on how much was spent on the food that was consumed. The ability to link such economic information as food purchasing, food consumption, and household socioeconomic characteristics to survey information on what consumers know about diet, health, good food preparation practices, and food safety issues is also lacking.

Data from government-sponsored national surveys, which are the primary sources available for policy and decision making, usually take a few years to collect and process. Yet for some policy-making decisions, data on the most recent market conditions are needed by the U.S. Department of Agriculture (USDA). Data on the supply of food, from growers and food manufacturers, are available on a relatively timely basis. However, with existing data sources, actual consumption of this food can only be inferred on an aggregate level. The most timely data on food purchases and food consumption are collected by market research firms to assess trends in food consumption. These data are collected for proprietary reasons, are of uncertain quality, and require payments, unlike data from government surveys, which are generally free to users. Because of the lack of access to timely data, the secretary of agriculture has less information for making some decisions affecting markets, programs, and the health of U.S. citizens than the executives of many companies in the private sector.

Interagency efforts have been undertaken to fill data gaps and develop comprehensive nutrition, health, and food consumption monitoring data systems. Most notably, in 1993, the Interagency Board on Nutrition Monitoring and Related Research (IBNMRR), chaired by senior officials of the U.S. Department of Health and Human Services and USDA, prepared a 10-year plan for a comprehensive nutrition monitoring and related research program. Mandated by the National Nutrition Monitoring and Related Research Act of 1990 (P.L. 101-445), the plan's goals were to further the collection of continuous, coordinated, timely, and reliable data by federal and state agencies; foster the use of comparable methods for collecting data and reporting results; promote related research; and disseminate and exchange information with data users. The IBNMRR published directories of federal and state nutrition-related datasets and monitoring activities and commissioned reports on the dietary and nutritional status of the U.S. population (see, e.g., Federation of American Societies for Experimental Biology, 1995; National Center for Health Statistics, www.cdc.gov/nchs/about/otheract/nutrishn/nutrishn.htm [June 2005]). The mandate for the board expired in 2003, although work remains to be done to achieve its goals. In addition, work is required to link other kinds of data, such as prices, to information on consumption choices and the consequences for diet and health.

PANEL CHARGE AND CONTEXT

In recognition of existing data gaps, the Economic Research Service (ERS) of the USDA received funding from Congress to improve the data infrastructure on food consumption and nutrition. As part of this effort, ERS asked the Committee on National Statistics of the National Academies to convene a panel of experts to review data needs to support research and decision making for USDA food and nutrition policies and programs. The panel was also charged to assess the adequacy of the current data infrastructure and to recommend enhancements to improve it. For both tasks, the panel was asked to consider improvements to the current data systems, rather than new data systems.

The primary basis for the panel's deliberations, given limited resources, was a workshop on Enhancing the Data Infrastructure in Support of Food and Nutrition Programs, Research, and Decision Making, which the panel convened on May 27-28, 2004. The workshop served as a forum for the USDA and other federal agencies with related policy responsibilities to discuss continuing and emerging data needs for policy and decision making, the data sources available to address these issues, and possible improvements or alternative data sources. During the workshop, representatives from six USDA agencies, the Food and Drug Administration (FDA), the National Institutes of Health (NIH), and the Environmental Protection Agency (EPA) discussed current and emerging data needs for policy and decision making related to food consumption. Representatives from the two key federal statistical agencies that produce food consumption and expenditure datasets, the National Center for Health Statistics (NCHS) and the Bureau of Labor Statistics (BLS), and representatives from private firms that produce data on food consumption and expenditures, the NPD Group and ACNielsen, discussed the strengths and limitations of their data. Outside researchers gave their reactions to the presentations and suggested possible improvements to the data infrastructure. The workshop summary is included as Appendix A.

This report is based on the discussions at the workshop and the deliberations of the panel. The report outlines key data that are needed to better address questions related to food consumption, diet, and health; discusses the available data and some limitations of those data; and offers recommendations for improvements in those data. The panel was charged to consider USDA data needs for policy making and the focus of the report is on those needs. It is important, however, to recognize that many policy issues and the data required to address them fall under the purview of other agencies with related policy missions. For example, EPA and FDA both share responsibility with the Food Safety and Inspection Service of USDA to ensure the safety of the nation's food supply. Likewise, several agencies within DHHS have missions related to the diet and nutritional adequacy of food consumed by Americans. This report does not explicitly cover the food consumption data needs of these other agencies, but it includes their needs when they overlap with those of the USDA.

The panel was charged to consider incremental changes in existing data systems that could be implemented (1) in a relatively short time frame, (2) at modest expense, (3) for general analytical use. A more comprehensive study would also consider data needs for assessing specific problems and population groups of interest, regardless of which agency was responsible and putting aside considerations of resources and time for development—as one example, the data requirements to fully understand the causes and consequences of obesity. Another example is the data requirements for assessing the health and nutrition of low-income people who receive (or are eligible for but do not receive) benefits from food assistance programs.[1]

The panel's study was conducted with the benefit of previous and continuing work of other studies from the National Academies. In the late 1990s, the Committee on National Statistics convened a workshop on evaluating food assistance programs in an era of welfare reform. The workshop participants considered data needs and research methods for assessing the effects of programs and changes in programs on household economic and food security and individual health and well-being (National Research Council, 1999). The workshop report discussed most of the relevant federal datasets on food and nutrition in some detail but did not make recommendations for changes in them. Workshop participants also discussed the utility of linking survey data with relevant administrative records, as well as the need for more state-level data.

In 2003 the Board on Agriculture and Natural Resources held a work-

---

[1] For this purpose, Logan, Fox, and Lin (2002) reviewed almost 100 data sources for their potential to support food assistance program outcomes research, identifying 13 data sources that are clearly useful, another 13 sources that could be useful if they were expanded in one or more ways, and 70 sources that are not useful because they are outdated, restricted to specific populations, lack sufficient content on program participation, nutrition, and health, or comprise administrative records that would be difficult to link to datasets with program participation information.

shop to consider ways to more efficiently and effectively conduct food and health research and education for promoting better health. In particular, the workshop considered how to begin to integrate relevant input from the agricultural and health sciences, although it did not explicitly consider the data infrastructure needs for integration (National Research Council, 2004).

The Institute of Medicine's Food and Nutrition Board and Board on Health Promotion and Disease Prevention recently completed a study to assess the factors responsible for the epidemic of obesity in children and identify promising approaches for prevention efforts. That study reviewed the scientific literature on the causes of childhood obesity and on obesity prevention programs and recommended a research and action agenda to assist in the prevention of obesity (Institute of Medicine, 2005). The study was not specifically charged to address improvements in the data infrastructure for evaluation of obesity prevention programs, but it did recommend that the federal government "strengthen support for relevant surveillance and monitoring efforts, particularly the National Health and Nutrition Examination Survey" (Institute of Medicine, 2005:6).

Finally, the Committee on National Statistics (CNSTAT) has under way a study of the measurement of food insecurity and hunger by the USDA; that measurement is obtained from a supplement to the December Current Population Survey (see Nord and Bickel, 2002; Nord, Andrews, and Carlson, 2004). The CNSTAT study recently completed a first phase report (National Research Council, 2005b) and will issue a final report that examines in depth the current food insecurity measure and possible alternatives to it. So as not to duplicate work, this report on improving data to analyze food and nutrition policies does not consider food insecurity measurement, although there is a brief discussion in Chapter 4 of the food insecurity scale and other food-related information in the Current Population Survey December supplement.

The rest of this chapter provides a context for consideration of data needs by briefly reviewing important factors that influence food purchasing and consumption behavior and the consequences for diet and health. It also provides a limited review of food consumption data needs for food safety policies and programs. Box 1-1 lists the acronyms used throughout the report. Table 1-1 summarizes key features of the major public and private surveys of data on food consumption, expenditures, and store sales that are considered in the report.

Chapters 2-4 examine the three sources that can provide data relevant to the panel's charge to examine the data infrastructure for food and

BOX 1-1
Acronyms Used in the Report

ACS	American Community Survey
APHIS	Animal and Plant Health Inspection Service, USDA
ATUS	American Time Use Survey
BEA	Bureau of Economic Analysis
BLS	Bureau of Labor Statistics
BMI	Body Mass Index
BRFSS	Behavioral Risk Factor Surveillance System
BSE	Bovine Spongiform Encephalopathy (mad cow disease)
CDC	Centers for Disease Control and Prevention, DHHS
CE	Consumer Expenditure Survey
CPS	Current Population Survey
CSFII	Continuing Survey of Food Intakes by Individuals
DHHS	U.S. Department of Health and Human Services
DHKS	Diet, Health, and Knowledge Survey
ECLS	Early Childhood Longitudinal Study
EPA	Environmental Protection Agency
FCBSM	Flexible Consumer Behavior Survey Module
FDA	Food and Drug Administration, DHHS
FSIS	Food Safety and Inspection Service, USDA
HHANES	Hispanic Health and Nutrition Examination Survey
HRS	Health and Retirement Study
IBNMRR	Interagency Board on Nutrition Monitoring and Related Research
IRI	Information Resources, Inc.
MEC	Mobile Examination Center, NHANES
MSA	Metropolitan Statistical Area
NET	National Eating Trends
NET	Nutrition and Education Training Program
NCHS	National Center for Health Statistics, DHHS
NHANES	National Health and Nutrition Examination Survey
NIH	National Institutes of Health
NIS	National Immunization Survey
OMB	Office of Management and Budget
PCE	Personal Consumption Expenditures
PDP	Pesticide Data Program
PSID	Panel Study of Income Dynamics
RDC	Research Data Center
SLAITS	State and Local Area Integrated Telephone Survey
UPC	Universal Product Code
USDA	U.S. Department of Agriculture
WIC	Special Supplemental Nutrition Program for Women, Infants, and Children

TABLE 1-1 Overview of Major Federal and Private-Sector Surveys on Food Consumption, Food Purchases by Consumers, and Food Sales in Stores

Type of Survey	Food Consumption (Intake) Surveys	
Survey	*National Eating Trends (NET)*	*National Health and Nutrition Examination Survey (NHANES)*
Sponsor	NPD Group	DHHS/USDA
Description	Food intake by individuals from 2-week diary; socioeconomic characteristics of all household members; where food purchased and eaten	Food intake by sampled persons from 24-hour recall for 2 nonconsecutive days beginning 2002 (first in-person, second by phone); household characteristics; food assistance program participation; socioeconomic characteristics, where food eaten, and detailed health measures from medical tests and examinations for sampled persons
Sample	Nationally representative; 2,000 households sent diaries each year	Nationally representative; some groups oversampled; 5,000 people examined each year
Frequency and Timeliness	Ongoing panel; 3-month lag between collection and release	Continuing since 1999; released approximately every 2 years with 2-year lag from collection; 2001-02 available (does not include day 2 of dietary intake)
Response Rate (approximate)	Not available	82 percent of eligible sample persons interviewed; 76 percent examined (1999-2000 round)

NOTE: For more information on NHANES, CSFII, and DHKS, see Chapter 2; for NET, see Chapter 3.

Discontinued Consumption Surveys	
Continuing Survey of Food Intakes by Individuals (CSFII) – Integrated with NHANES in 2002	*Diet and Health Knowledge Survey (DHKS)* – Source of questions for new Flexible Consumer Behavior Survey Module in NHANES
USDA	USDA
Food intake by sampled persons from 24-hour recall for 2 nonconsecutive days (in-person, 1994-1996 round); household characteristics; food assistance program participation; last month's food expenditures; socioeconomic characteristics of members aged 15 and older; where food eaten and health measures for sampled persons	Supplement to CSFII in 1989-91 and 1994-96; questions about diet and health knowledge and attitudes, use of food labels, factors in shopping, food preparation practices
Nationally representative; oversampling of low-income people; 5,000 people per year over 3 years	One adult aged 20 and older who completed a dietary intake from each household in CSFII sample
Conducted most recently in 1989-91, 1994-96, and 1998 (children aged 0-9 only); most questions now in NHANES	Conducted as supplement to CSFII in 1989-91 and 1994-96; not currently available
80 percent of sampled household members completed first day of dietary intake; 76 percent completed second day (1994-1996 round)	74 percent of eligible adults (see "Sample" above)

continued

TABLE 1-1 Continued

Type of Survey	Food Purchases by Consumers	
Survey	*Consumer Expenditure Survey (CE)*	*Combined Outlet Consumer Panel*
Source	BLS	Information Resources, Inc. (IRI)
Description	*Household Survey:* household characteristics; socioeconomic characteristics for members aged 15 and older; household food assistance benefits; usual monthly or weekly expenditures on food by type of outlet *Diary Survey:* Usual weekly food expenditures; price of purchased food by type and outlet	Household panel members scan their food purchases from retail outlets; includes prices, quantities, promotion information, and demographics
Sample	Nationally representative; 7,500 consumer units per year in Household Survey (5 quarters of information); 7,500 households per year in Diary Survey (two 1-week diaries)	Nationally representative; panel of 50,000 households
Frequency and Timeliness	Continuing since 1980; released annually with 1-year lag from collection, 2003 available	Monthly data with 12-day lag between collection and release
Response Rate (approximate)	78 percent, 2001 Household Survey; 75 percent, 2001 Diary Survey	Not available

NOTES: For more information, see Table 2-2 in Chapter 2 on the CE; see Chapter 3 on the Combined Outlet Consumer Panel, HOMESCAN Consumer Panel, and CREST; see Chapter 4 on other surveys that include limited data on food purchases.

INTRODUCTION 17

Consumer Report on Eating Share Trends (CREST)	*HOMESCAN Consumer Panel*
NPD Group	ACNielsen
Prepared food purchases by individuals at commercial restaurants and other outlets, including fast food outlets; includes prices; identifies outlets	Household panel members scan their food purchases from retail outlets; includes prices, quantities, and promotion information for items with UPC codes; item identification and weights for items lacking UPC codes; and demographics
Online survey, weighted to be nationally representative; 3,000 adults and 500 teenagers daily	Nationally representative; panel of 61,500 households (only one-quarter report both UPC and non-UPC purchases)
3-month lag between collection and release	Monthly data with 3-week lag between collection and release
40 percent	85 percent

continued

TABLE 1-1 Continued

Type of Survey	Food Sales in Stores	
Survey	*Custom Store Tracking*	*Scantrack Services*
Source	Information Resources, Inc. (IRI)	ACNielsen
Description	Point-of-sale data for food stores, food/drug combinations, and mass merchandisers	Point-of-sale data for food stores, food/drug combinations, and mass merchandisers
Sample	Nationally representative; 32,000 retail outlets across the U.S.	Nationally representative; 4,800 stores representing more than 800 retailers in 52 major markets
Frequency and Timeliness	Monthly data with 12-day lag between collection and release	Monthly data with 10-day lag between collection and release
Response Rate (approximate)	Not applicable	Not applicable

NOTES: See Chapter 3 for discussion of Custom Store Tracking and Scantrack Services.

nutrition research and policy: federal datasets on food and nutrition (Chapter 2); proprietary sources of food consumption and expenditure data (Chapter 3); and federal datasets that might provide some of the otherwise missing data (Chapter 4). These chapters also consider the limitations of the various data sources in addressing questions about food consumption patterns. Chapter 5 presents the panel's recommendations.

FOOD CONSUMPTION DECISIONS

The question of "What shall we do for dinner tonight?" has probably occupied more time and thought and generated more tension than most people are willing to admit. The process of answering this question usually goes something like this: What are you hungry for? What food do we have at home? Do we have the time or energy to cook? Will the kids eat it? Can we afford to go out? Will the grocery store or restaurant be crowded at this time of day and do we have enough time to stop there? What would be healthy to eat or do we care about that right now?

It is clear that a number of factors go into making this daily decision. They include not only individual- and household-level factors (such as income, time resources, knowledge, skills, and preferences), but also factors outside of the household (prices and the availability of stores and restaurants). At perhaps a lower level of consciousness, the larger policy and media environment probably also play a role. For example, concern over mad cow disease may (or may not) trump a craving for the big juicy burger and fries just advertised on a television commercial. The purpose of considering the context of food decisions is to highlight the types of data that need to be collected in order to understand food consumption choices and to address the questions about food consumption that are listed above.

In making decisions regarding food consumption, households and individuals consider their resource levels. These resources include monetary resources (income and asset levels), which are not always adequate for food consumption. Evidence from the current USDA measurement of food insecurity indicates that people in as many as 11 percent of U.S. households in 2003 worried about their ability to obtain adequate food for the family and were not always able to do so due to economic deficits (Nord, Andrews, and Carlson, 2004:3). Resources also include time—the amount of time available for food preparation and eating and for other activities. In 2003, Americans spent an average of 1.2 hours per day on food consumption and 0.5 hours per day on food preparation and cleanup (Bureau of Labor Statistics, 2004c:Table 1). These averages mask differences in time constraints for food preparation and consumption among different kinds of households, such as a family with two working adults compared to a family with one adult working outside and one inside the home or to a family with only one adult. Household members may also have a set of skills or informational resources available—for example, information on which foods are healthy and food preparation skills. Finally, individual household members have different food preferences (and allergies or aversions to some foods).

There are other factors that contribute to food consumption decisions. The amount and types of foods that can be consumed given a household's resources depend on the prices households face and the availability of different types of food (for example, the presence and types of grocery stores, restaurants, and food retailers and the variety of foods they carry). Packaging may also affect food consumption decisions, as may labeling that identifies the ingredients and nutrient elements that individuals may or may not prefer to eat or may perceive as harmful. Technology for food production and preparation is also a factor for households, both for their own use and for the production of food away from home.

Some theories about why low-income populations have a higher prevalence of being overweight or obese focus on the contextual factors. For instance, one argument is that the lack of major grocery stores and health food stores in low-income neighborhoods contributes to higher prices and lower availability of healthier foods, such as fruits and vegetables, while convenience stores and fast-food restaurants are plentiful in these neighborhoods. In other words, tasty, energy-dense, and low-micronutrient foods are readily available at low cost. These foods may be a logical choice if healthful foods are difficult to obtain and if a family must be fed with limited economic resources. However, research shows a more complex picture. For example, in a 1996 survey of low-income households, most people were able to shop at supermarkets within or close to their neighborhood, though one-third traveled more than 4 miles to shop for food, most often citing high prices and lack of stores in their neighborhood as reasons (Ohls et al., 1999:xiii-xiv; see also Cole, 1997).

Changes in technology have also been examined as possible contributors to consumption trends (Lakdawalla and Philipson, 2002). Specifically, it is hypothesized that technological advances in food preparation have reduced time costs and increased the quantities and varieties of foods that are produced both in the home and by mass producers (Cutler et al., 2003). Some of these advances, however, have likely contributed to less healthy eating habits.

Households also make decisions within a policy environment. Policies directly related to food consumption include food and nutrition assistance and education programs, most of which are under USDA, as well as food standards. Food assistance programs include the Food Stamp Program, the National School Lunch and Breakfast Programs, the Special Supplemental Nutrition Program for Women, Infants, and Children (WIC), and several smaller programs, including the Child and Adult Care Food Program, the Commodity Supplemental Food Program, the Emergency Food Assistance Program (TEFAP), the Food Distribution Program on Indian Reservations, Meals on Wheels for the elderly (a DHHS program), the Special Milk Program, the Summer Food Service Program, and the WIC Farmers'

Market Nutrition Program. Food education programs include Food Stamp Nutrition Education, the Nutrition Education and Training (NET) Program, Team Nutrition for School Meals, and the nutrition component of the WIC Program. Food standards include sugar restrictions on cereals in the WIC Program and nutrition standards for school meals. (See also Institute of Medicine, 2004, on criteria for selecting WIC food packages.)

Policies toward the marketing and labeling of food, such as nutrition labeling guidelines and guidelines for health claims of foods, may also contribute to consumption decisions. Public education campaigns for healthful eating, such as the Food Guide Pyramid, and on food safety may also have an impact on the foods that Americans eat and the methods they use to prepare them. In addition, medical education standards and health care delivery policies and practices regarding nutrition and diet may affect decisions that health care consumers make about eating. Finally, food safety policies, including federal and state regulations for food production and preparation, inspections, and surveillance of food-borne illness, affect the quality and variety of foods that reach the consumer.

This overview makes it clear that a great deal of information is needed to fully understand food consumption decisions and their consequences for diet and health. In addition to data on what foods people eat and what they prefer to eat, information on household resources—income, assets, time, education, health and diet knowledge, and food preparation skills—is needed. Environmental-level information is also needed—that is, information on prices of food and related goods; availability of different foods; availability of grocery stores, food retailers, and restaurants; and marketing practices (such as amount of advertising exposure, target audiences, coupons or other incentives, packaging and display). Finally, information on policy interventions, such as food stamps and other public assistance programs, government-assisted marketing efforts (such as the Dairy Board, over which USDA has oversight), labeling regulations, and public health initiatives related to diet is also needed.

FOOD SAFETY ISSUES

A complete assessment of data needs to support food safety policy is beyond the scope of this panel's report, but a few comments regarding the current state of data to support policy analysis are important for context.

The main point of interface between this panel's charge and food safety policy is the use of food consumption data to support exposure assessment for food safety hazards.

Dramatic changes in what food is eaten and how food is prepared during the last few decades have altered the type and incidence of food safety risks for U.S. consumers. For example, over the past 40 years, Americans markedly increased the percentage of their total food dollar that was spent on food eaten away from home (from 29 percent in 1963 to 47 percent in 2003), as well as the percentage of their dollar for food eaten away from home that was spent at fast-food outlets (from 26 percent in 1960 to 38 percent in 2000). Since 1970, annual consumption of red meat has dropped by 18 pounds, while annual consumption of chicken has increased by 37 pounds (see www.ers.usda.gov/Briefing [June 2005]).

These kinds of changes mean that exposure assessment in the future—whether to food-borne pathogens or pesticide residues—will need to be based on the most current food consumption patterns and should include information about methods of preparation, including consumption of undercooked, raw, or unwashed foods, and whether prepared at home or obtained from a retail outlet. Ideally, such data should provide sufficient detail to distinguish populations at particular risk for some hazards, such as expectant mothers, young children, the elderly, and the immunocompromised.

Food-Borne Pathogens

It is now widely recognized that microbial food-borne pathogens are the most important food-borne hazard. Food-borne pathogens continue to evolve and adapt, with such new hazards emerging as *Salmonella enteritidis* in eggs, *E. coli* O157:H7 in ground beef, and the potential for BSE prions in beef (National Research Council, 2003). Exposure assessment is still in its infancy for these hazards. In food-borne illness outbreaks, the food source is often unidentified. Furthermore, there is imperfect understanding of the dose-response relationship for many pathogens, and attempts to estimate these relationships are not well developed.

A risk assessment by the USDA Food Safety and Inspection Service (FSIS) for *E. coli* O157:H7 used data from multiple sources for risk characterization and exposure assessment. These sources included: reported illnesses and food sources from the 10 surveillance sites in the FoodNet system (located in California, Colorado, Connecticut, Georgia, Maryland,

INTRODUCTION   23

Minnesota, New Mexico, New York, Oregon, and Tennessee), which is run cooperatively by the Centers for Disease Control and Prevention, the FDA, and the FSIS; the incidence of the pathogen in the food supply at various points along the supply chain from data provided by the FSIS and the USDA Animal and Plant Health Inspection Service (APHIS); and intake data regarding ground beef consumption (U.S. Department of Agriculture, 2001).

The ability to link data from different sources for such exposure assessments, however, is far from ideal. FoodNet monitors illness outbreaks in its 10 sites or catchment areas by obtaining information on the incidence of different food-borne pathogens at laboratories in these areas, but the laboratory data may not indicate the exact food product source. (FoodNet validates outbreak data by reviewing the methods and tests used by the FoodNet laboratories and surveying physicians in the catchment areas to determine under what circumstances they order samples to be analyzed by these laboratories.) The FSIS monitors food-borne pathogens in meat and poultry slaughter and processing plants, but it does not make these data public, except for periodic summaries. The APHIS monitors pathogen incidence in animals on farms, but not on a regular basis for all pathogens of potential human health interest.

More importantly, there is no systematic monitoring of food-borne pathogens in the food supply at the retail and household level. The National Health and Nutrition Examination Survey (NHANES) obtains much relevant data for exposure assessment, including 2-day dietary recall begun in 2002. However, the survey does not currently obtain information on food preparation (for example, washing raw foods) and other topics thought to be most useful for this purpose (see Chapter 2).

The FDA periodically carries out a random-digit dialing Food Safety Survey of 2,000-4,000 households to monitor perceptions of individual and societal risk related to food consumption, food-handling practices in home-prepared food, understanding and use of food product safety labels, food allergies, consumption of potentially risky foods, attitudes toward new food technologies, perception, knowledge, and experience of food-borne illness, and sources of food safety knowledge. However, the Food Safety Survey, which was fielded in 1988, 1993, 1998, and 2001, does not include data on actual food consumption, its data on household characteristics are limited, and its response rates are low (61 percent in 2001—see www.cfsan.fda.gov/~lrd/ab-foodb.html [June 2005]).

FDA has conducted periodically since 1982 the Health and Diet

Survey, which addresses more broadly the health-related knowledge and attitudes of people aged 18 and over in households. Last conducted in 2004, it has been used to study people's awareness of relationships between diet and risk for chronic disease, consumer use of food labels, weight loss practices, and the effectiveness of the National Cholesterol Education Program. Like the Food Safety Survey, the Health and Diet Survey does not collect data on food consumption or economic characteristics of respondents' households, and its response rates are very low (41 percent in 2002—see www.cfsan.fda.gov/~1rd/ab-nutri.html [June 2005]).

FoodNet includes a population survey component in which residents of its catchment areas are contacted and asked about recent diarrheal disease, treatment sought, and whether foods causing known outbreaks of food-borne illness have been consumed. However, the FoodNet population information does not include questions about food preparation and storage or residents' knowledge of food safety issues, and information on household characteristics is limited (www.cdc.gov/foodnet/what_is.htm [June 2005]).

The gaps in the available data mean that heroic assumptions are required to link together pathogen incidence data, intake data related to specific subpopulations, and food preparation data in order to carry out exposure assessments for one or more food-borne pathogens. Attention to filling these gaps to improve the validity and reliability of such linkages will be important for future food safety risk assessment and policy analysis.

Pesticide Residues

Residues from pesticides in food can also be a problem, although it is not clear which kinds of residues are currently the most problematic or may become so in the future. Exposure assessments for pesticides currently use a methodology developed by the EPA to carry out its mandate under the 1996 Food Quality Protection Act to periodically review tolerance limits and reregister pesticides for compliance with updated standards. The Act required EPA to give special consideration to children's exposure and exposure to multiple residues with similar toxicity. EPA's exposure assessment process first uses a conservative assumption of residues equal to maximum residue limits, such that data on actual residues are not required. When this initial assessment indicates a potential risk from a particular pesticide, EPA refines its assessment using realistic exposure data.

Actual pesticide residue data are collected through the USDA's Pesticide Data Program (PDP) and the FDA's Total Diet Study. The PDP began in

INTRODUCTION

1991. It operates in 10 states and is designed to capture information on actual residues in the food supply as close as possible to when food is actually eaten. The Total Diet Study began in 1961. It obtains samples of food purchased by FDA personnel in selected cities, which are then analyzed by FDA laboratories; the results are used to estimate exposures by weighting to food consumption patterns from the Continuing Survey of Food Intakes by Individuals (the CSFII food consumption data are now part of NHANES—see Chapter 2).

The PDP and the Total Diet Study represent important sources of information about the incidence of residues in the food supply over time. They have been used, for example, to assess actual risks as a way of better understanding outcomes of pesticide regulation (Day et al., 1995; Kuchler et al., 1997).

Evaluating Consumer Education

One more point on food safety information is worth noting. Data on consumer food safety knowledge and practice would be crucial for developing any consumer education or labeling efforts aimed at safety. Questions in these areas could be part of the addition of a health knowledge component to the NHANES that ERS is developing and that we support (see Chapters 2 and 5). They could also be added to the FoodNet population survey component. Conversely, more detailed household characteristics, as well as some information on food consumption, could be useful to add to the FDA Food Safety and Health and Diet Surveys, perhaps asking questions of subsamples to reduce burden on respondents.

2

Federal Datasets on Food and Nutrition

This chapter reviews the primary sources of data from federal surveys on food consumption, food expenditures, and dietary attitudes and knowledge. Those surveys are the National Health and Nutrition Examination Survey, the Continuing Survey of Food Intakes by Individuals, which was incorporated into NHANES beginning in 2002, and the Consumer Expenditure Survey. Table 2-1 summarizes the design and lists the content relevant to food and nutrition policy and research of these three surveys and the Diet and Health Knowledge Survey that was included in two rounds of the Continuing Survey of Food Intakes by Individuals. The periodic Health and Diet Survey discussed in Chapter 1 also provides information on adults' health-related knowledge and attitudes.

NATIONAL HEALTH AND NUTRITION EXAMINATION SURVEY

The continuing National Health and Nutrition Examination Survey (NHANES) is conducted by the National Center for Health Statistics (NCHS) of the Centers for Disease Control and Prevention in the U.S. Department of Health and Human Services. NHANES monitors the health and dietary practices and outcomes of Americans and is used in developing public health policy. The dietary component of NHANES, a cooperative effort of the Departments of Agriculture and Health and Human Services, is called "What We Eat in America" (www.barc.usda.gov/bhnrc/foodsurvey/wweia.html [June 2005]). NCHS is responsible for the sample design and data collection for the dietary component; USDA is responsible for the design of the dietary data collection procedure, maintenance of the databases that are used to code and process the dietary intake information, and review and processing of the dietary information (see "Integrated NHANES and CSFII" below).

The NHANES has evolved from a periodic survey that originated in 1971, when the National Health Examination Survey was combined with the National Nutrition Surveillance System, to a continuously fielded survey. NHANES I data were collected for 1971-1975, NHANES II for 1976-1980, and NHANES III for 1988-1994. There was also an Hispanic Health and Nutrition Examination Survey (HHANES), which was conducted from 1982 through 1984. In 1999, NHANES became an ongoing survey, with detailed health, nutrition, and medical information collected from about 5,000 participants annually. Beginning with data for 1999-2000, NHANES findings have been released at 2-year intervals. To enable users to produce estimates from the NHANES public-use microdata files with sufficient reliability, sample weights are provided on a pooled basis for each 2 years' worth of information. To protect confidentiality of respondents, not all variables are included in the public-use microdata files.

The sampling process is stratified and multistage: counties or groups of contiguous small counties are designated as primary sampling units (PSUs), and each year 15 PSUs are selected into the sample for household visits. Within PSUs, blocks or groups of blocks are selected, then households, and, finally, one or more individuals within households. The sample does not include people living in institutions or members of the Armed Forces. At sampled households, interviewers obtain the demographic characteristics of all household members, and one or more household members are selected for interview and examination by using fixed sampling fractions that distribute the sample into specific age-sex-race-ethnicity-income categories. If a child under the age of 6 is selected into the sample, then a proxy interview is conducted with the child's primary caretaker. Interviews with children aged 6-11 years old are conducted through an assisted interview with a caretaker present. Blacks, Mexican-Americans, 12- to 19-year-olds, people aged 60 and over, pregnant women, and (beginning in 2000) people in low-income households are oversampled (see www.cdc.gov/nchs/data/nhanes/guidelines1.pdf [June 2005]).

In addition to demographic and economic background information and self-reported health status for each individual in the sample, physiological information, including precise measurements of height and weight

TABLE 2-1 Design Features and Relevant Content for Food and Nutrition Research of Four Surveys: the National Health and Nutrition Examination Survey (NHANES), the Continuing Survey of Food Intakes by Individuals (CSFII), the Diet and Health Knowledge Survey (DHKS), and the Consumer Expenditure Survey (CE)

Feature	NHANES	CSFII
Time Period	Continuing since 1999; released at approximately 2-year intervals, with data pooled for 2 years for reliability	Conducted most recently in 1989-1991, 1994-1996, and 1998 (children aged 0-9 only); folded into NHANES in 2002
Universe	Civilian noninstitutionalized population	1994-1996: Same as NHANES
Design	Each year about 11,000 households screened in 15 primary sampling units (counties); about 3,000 households identified with 6,000 eligible people, of whom 5,000-5,500 interviewed and 4,600-5,200 examined in mobile exam centers (MEC) Oversampling of blacks, Mexican-Americans, low-income people not black or Mexican (beginning in 2000), people aged 12-19 or aged 60 and over, pregnant women	1994-1996: 15,000 people in households (5,000 people per year) provided dietary intake information Oversampling of low-income people
Major Questionnaire Components	Household screeners; Family questionnaire; Sample person questionnaire; MEC audio computer-assisted self interview; MEC computer-assisted personal interview; MEC dietary recall; MEC examination; MEC laboratory analysis	Household interview; First-day in-person dietary recall; Second-day in-person dietary recall

DHKS	CE
Conducted as a supplement to CSFII in 1989-1991 and 1994-1996; not now carried over to NHANES	Continuing since 1980; released yearly; two components, *Diary Survey* and *Household Survey*
Same as NHANES	Same as NHANES
Random sample of adults aged 20 and over with a completed dietary intake for day one, who were interviewed by telephone follow-up 2-3 weeks after the second day of dietary intake	*Household Survey*: 7,500 consumer units per year; each month, one-fifth of sample is new; households in sample for five quarterly in-person interviews; respondent is anyone aged 16 or older who knows household finances *Diary Survey*: 7,500 consumer units per year, each of which fills out two consecutive weekly diaries No oversampling
Knowledge questions; Attitude questions; Factors related to grocery shopping; Food label questions; Behavior questions; Food safety questions	*Household Survey*: Demographic characteristics; Work experience; Expenditures by month (65% of items); Usual expenditures per quarter (35% of items); Real assets; Financial assets; Last 12-months' income; Taxes *Diary Survey*: Demographic characteristics; Work experience; Income; Taxes; First week diary; Second week diary

continued

TABLE 2-1 Continued

Feature	NHANES	CSFII
Household Background Characteristics	Family size, ages, relationships (data collected for sampling but not publicly released); health insurance; housing characteristics; pesticide use; smoking	Family size; tenancy
Person Background Characteristics	Sampled persons: Age, race, sex; country of birth, marital status, early childhood; education; occupation; social support	All household members: Age, race, sex Members aged 15 and over: Education, employment status, occupation, hours worked last week, usual hours worked, reason not working
Income	Total household income in last 12 months; who received different income types in last 12 months (number of months for welfare assistance)	Total household income last year; last month's income by source; savings or cash assets under $5,000
Food Assistance Program Participation	Household and sampled person food stamp and WIC participation in last 12 months, number months receiving food stamps and WIC; amount food stamps received by household last month; sampled person participation in school breakfast and lunch (usual times per week), Meals on Wheels in past 12 months, and summer program meals	Household food stamp participation in last 12 months, value of food stamps, when last received, members eligible now; each member's WIC participation and how long; school-age children's school breakfast and lunch participation; younger children's child care feeding participation

DHKS	CE
Obtained from main CFSII	*Household Survey*: Family size; health insurance; inventory of housing and financial assets, durable goods (first quarter) *Diary Survey:* Family size; housing, vehicle ownership
Obtained from main CFSII	*Household Survey*: All household members: Age, race, sex, marital status, education; Members aged 14 and over: Work experience, job characteristics *Diary Survey:* Same as *Household*
Obtained from main CSFII	*Household Survey*: Members aged 14 and over: Earnings and retirement benefits (each quarter); Income by source, prior 12 months (2nd and 5th interviews) *Diary Survey:* Members aged 14 and over: Earnings and retirement benefits Household: Other income by source, prior 12 months
Obtained from main CSFII	*Household Survey*: Benefits from food stamps (and months received); value of other free meals (each quarter) *Diary Survey:* Same as *Household*

continued

TABLE 2-1 Continued

Feature	NHANES	CSFII
Food Expenditures	None	Usual weekly or monthly household expenditures at grocery stores (total and on food items), at specialty stores, at fast food or carryout places (for food brought into the home and food bought and eaten away from home)
Food Eating and Shopping Practices	Sampled persons: Where food obtained; number of times eat at restaurant in a week	Sampled persons: Where food obtained (store, restaurant, fast food place, etc.); whether eaten at home or away; sources of water for drinking, cooking, preparing beverages; time of eating; name of eating occasion (e.g., breakfast); who in household does planning, shopping, and preparing meals; how often shop and type of store
Food Security	18-item household food security scale module; individual-level questions for children, adolescents, and adults	Food sufficiency indicator

DHKS	CE
Obtained from main CSFII	*Household Survey,* Usual weekly expenditures at supermarkets, specialty food stores, school meals; usual monthly expenditures for liquor at home, liquor away from home, food away from home (every quarter) *Diary Survey:* Usual weekly expenditures at supermarkets, specialty food stores (last month); From the diary: Price of each purchase of food away from home, categorized by fast food-type outlets, full-service meals, vending machines, employer and school cafeterias, catered affairs; price of each item purchased for consumption at home, categorized by grain products, bakery products, beef, pork, poultry, other meats, fish and seafood, fats and dressings, eggs and dairy products, fruits and juices, sugars, vegetables, other food items, nonalcoholic drinks, alcoholic drinks
Obtained from main CSFII	Where food obtained (see "Food Expenditures" above)
Obtained from main CSFII	None

continued

TABLE 2-1 Continued

Feature	NHANES	CSFII
Person Health and Nutrition Characteristics	Sampled persons: Alcohol use; balance; biochemistries and hematologies; blood lipids; blood pressure; blood and urine analysis; body composition; bone markers; dermatology; diabetes; heart disease; illegal and prescription drug use; immunization status; kidney disease; medical conditions; oral health; osteoporosis; pain; physical activity; pregnancy status; respiratory health; self-assessed health status; sexual behavior; smoking; vision; weight and height measures; weight history	Sampled persons: Alcohol use; food allergies; physician-diagnosed diseases; physical activity; hours of TV or videos watched yesterday; pregnancy or lactation status; self-assessed height and weight; self-assessed health status; smoking
Person Dietary Intake	Sampled persons: 2-day 24-hour dietary recall of foods and amounts eaten (first day at the MEC; second, nonconsecutive day by telephone); all dietary supplements (e.g., vitamins, herbals—name, strength, days and amount in past 30 days, dosage, length of use); medications; water intake; salt use; special diet; milk consumption history; frequency of consumption of milk, green leafy vegetables, legumes, fish, shellfish; comprehensive food frequency questionnaire	Sampled persons: 2-day 24-hour dietary recall of foods and amounts eaten (first day in-person; second day 3-10 days later in-person); dietary supplements (categories); water intake; salt use (specific for certain foods); special diet

FEDERAL DATASETS ON FOOD AND NUTRITION

DHKS	CE
Obtained from main CSFII	None
Obtained from main CSFII	None

continued

TABLE 2-1 Continued

Feature	NHANES	CSFII
Person Diet and Health Knowledge	None (except for food frequency questions above)	Obtained from DHKS

NOTE: Information pertains to most recent version of each survey: Integrated NHANES/CSFII (beginning in 2002); 1994-1996 CSFII; 1994-1996 DHKS; current CE. A Flexible Consumer Behavior Survey Module will be fully implemented in

and blood chemistry, is obtained. This information is collected when respondents visit mobile examination centers (MECs), which travel around the country to administer the survey. Analysis of blood chemistry includes measures of total cholesterol and HDL cholesterol. Respondents are also asked about physical activities and, beginning in 2003-2004, those aged 6 and over are asked to wear a physical activity monitor for 7 days and return it by mail.

With regard to food consumption, respondents are asked to complete a 24-hour dietary recall for two nonsuccessive days; the first recall is conducted in person at the MEC and the second by telephone. Prior to 2002, the dietary recall covered only one 24-hour period, except that NHANES III (1988-1994) collected a second dietary recall in person from a 5-10

DHKS	CE
Sampled adults: Knowledge of recommended food group servings, healthiness of own diet; knowledge of foods with fats and cholesterol and whether high or low amount; self-assessment of weight; perceived importance of dietary guidelines; agreement with beliefs relevant to dietary behavior; importance for shopping of safety, nutrition, price, how easy to prepare and keep, taste; understanding of various aspects of food labels, perceived usefulness of labels; frequency of using lower-fat items, adding fat to vegetables, eating fried chicken, bakery products and chips, beef, pork or lamb, eggs; frequency of washing fresh produce, peeling fruit, eating vegetable peels	None

NHANES in 2007-2008, with questions drawn from the DHKS and CSFII together with new questions (see text); some parts of the DHKS were asked in NHANES in 2005-2006.

percent subsample of participants. In 2003-2004, NHANES included a detailed food frequency (propensity) questionnaire as well, which was originally developed by the National Cancer Institute (http://riskfactor.cancer.gov/prs/abstracts/diet2.html [June 2005]). Since 1999, NHANES has administered the same food security questions that are included in a supplement to the December Current Population Survey and used by USDA to estimate the percentage of U.S. households that are food insecure and the percentage that experience food insecurity with hunger.[1]

---

[1] NHANES includes the 18 questions that are used to form the food insecurity scale, but not the additional questions on food expenditures and ways of coping with not having enough food that are part of the CPS supplement (see National Research Council, 2005b: Box 2-1; App. A).

The two dietary recalls for all participants that have been collected in NHANES since 2002 allow an estimate of intra-individual variance in nutrient intake, which can be used to adjust the observed population variance to reflect the day-to-day differences in nutrient consumption of individuals. Similar adjustment of the population variance in foods consumed (e.g., vegetables) for 2003-2004 will be based on the food propensity questions.

For accurate estimates of nutrient adequacy in the population, obtaining at least 2 days of food intake for sampled individuals is essential. While only a few days of food intake information are not sufficient to estimate the usual nutrient intake of an individual, they will support estimates for the population as a whole and for many groups of interest.

CONTINUING SURVEY OF FOOD INTAKES BY INDIVIDUALS

Prior to 2002 the Agricultural Research Service of the USDA sponsored periodic rounds of the Continuing Survey of Food Intakes by Individuals (CSFII). This in-person survey of approximately 5,500 individuals per year contained information about food consumption, which was collected from two days of dietary recall (in the 1994-1996 round). The survey also collected information on food expenditures and shopping practices, such as how often and where food shopping was conducted, how much money was spent in the past 3 months on types of food and grocery items, and the usual amount of money spent on food away from home and carryout and fast food. Information on participation in food assistance programs and food sufficiency was also collected. Data from the CSFII were used by the USDA to develop the thrifty food plan and the first food guide pyramid and to conduct risk assessment analyses for food safety issues. They were also used by the Environmental Protection Agency (EPA) and the Food and Drug Administration (FDA) to conduct exposure assessments for pesticides and microbials.

A telephone supplement to the CSFII for the 1989-1991 and 1994-1996 rounds was the Diet and Health Knowledge Survey (DHKS). This survey obtained information from adult CSFII respondents who completed at least the first day of dietary intake about their views and knowledge of a healthy diet and nutrition, how their diets compared to a healthy diet, what they understood about nutritional labeling and their use of labeling in making food purchase decisions, and their food preparation practices (U.S. Department of Agriculture, 2000).

INTEGRATED NHANES AND CSFII

To improve efficiencies of data collection, the CSFII was largely integrated into the continuing NHANES beginning with the 2002 collection. It is important to note that the two surveys have different focuses, and there are some differences in the collection of food data in the NHANES and the last-used version of the CSFII. NHANES focuses on health and nutrition, while CSFII focused solely on the kinds and amounts of foods eaten and did not include a medical examination. CSFII collected dietary recall data for two days from every person in the sample. As indicated above, NHANES began collecting two days of dietary recall from the entire sample of survey participants in 2002. A key difference in content is that the CSFII asked questions regarding food shopping, spending, and preparation, as well as the DHKS component described above, but none of these questions were carried over to NHANES.

The integrated NHANES and CSFII provides extensive information on health and medical conditions and represents a valuable resource for improving understanding of the relationships between diet and health. It will certainly become widely used as it is the only source of information on actual food consumption. There are, however, limitations in using the survey to address some critical policy questions.

Limitations

Exclusion of Some CSFII Content

NHANES does not currently collect information on household food expenditures, food shopping patterns, or diet and health knowledge, all of which had been collected by the main CSFII or the DHKS supplement.[2] These data have previously been used to understand the link between nutrition, diet, and health knowledge and actual consumption of food. They have also been used to analyze the effects of food labeling practices and policies on consumption. Expenditure information has been used to relate food expenditures to food consumption patterns across different population groups and for those who participate in food assistance pro-

---

[2]Plans are under way to add questions on these topics to NHANES—see "ERS Supplement Initiative" below.

grams: for example, to study the relationship between expenditures on carry-out and fast food outlets as a percentage of total budget and diet quality. Studies have indicated that increased use of fast food outlets is associated with increased energy intake, higher fat content, and lower intake of fruits and vegetables (Chou, Grossman, and Saffer, 2004; French et al., 2001; French, Harnak, and Jeffey, 2000).

Limited Economic Data

Economic analyses of consumer choices rely heavily on information on budget constraints of families. These analyses require information on the resources available to families—that is, income and asset levels and time resources. Analyses of food assistance and other public assistance programs also require information on income and assets to determine program eligibility, as well as information on the benefits households receive from these programs. NHANES collects some, but not extensive, data on income, assets, and receipt of public assistance. The survey asks respondents whether members of their households received income or public assistance benefits from various sources and to report a combined total level of income. Data on the amount of income or benefits from specific sources are not collected, although beginning in 2005 last month's food stamp benefit amount is being collected. Very little information on assets is collected—the survey contains only a question on home ownership status and a question on receipt of interest income from financial assets.

Design Limitations

The NHANES sample size of 5,000 people per year is sufficient for reporting mean dietary intakes or the prevalence of overweight and other frequently occurring health conditions and for many other purposes. However, the sample size is small for reporting on subgroups or conditions that are not frequently found in the population. The sample size is also small for multivariate analyses of the relationships of economic and social factors to food consumption and nutrition and health outcomes. In addition, NHANES has a highly clustered sample design with different sampling rates for population groups. These features generally increase the sampling errors of estimates compared with a simple random design and result in large variations in sampling errors for different analytical variables. As a consequence, the effective sample size for many analyses, in terms of sampling error, is smaller than the number of individual cases. Indeed, when the 2001-2002 NHANES data became available, users were advised to pool the data over 4 years covering 1999-2002 ("NHANES Analytical Guidelines—June 2004," at www.cdc.gov/nchs/nhanes.htm [June 2005]).

The highly clustered design of NHANES could lead to some biases in estimating consumption of foods that are very localized in production and transport. For earlier rounds of NHANES, it was suggested that the scheduling for the mobile examination centers could bias estimates for foods consumed at different times of the year: to minimize road delays and accidents, the MECs typically visited sampled areas in the north in the summer and in the south in the winter. However, under the continuing design, the three sets of MECs that cover the 15 areas in the sample each year are tightly scheduled in a manner that reduces the possibility of seasonal bias.

Another problem that may affect the dietary intake data is that the number of recalls is not evenly spread over the days of the week. Depending on when they had their first interview, respondents are randomly assigned to a day of the week to visit the MEC, where they are asked to tell interviewers about their food consumption for the previous 24 hours. But interviews sometimes have to be rescheduled. To the extent that weighting does not equalize the intake data by day of the week, then estimates of average intakes may be affected because of differences in daily consumption patterns. In particular, food consumption on Fridays, Saturdays, and Sundays differs from consumption during other days of the week. In a study of weekend eating, Haines et al. (2003) found a substantial increase in energy, fat, and alcohol intake on weekend days over other days of the week—as much as 115 calories per day in the 19- to 50-year-old age group.

One weakness of the complicated structure of the NHANES survey is that it takes time to process and edit the data before they are released publicly: data are usually not released until about two years after they were collected. Since NHANES became a continuous survey and processes for data production have become streamlined, it may take less than two years to edit and release the data in the future.

Potential for Improved Data

With increasing reliance on the continuing NHANES as the only large national survey with detailed data on food consumption, there will be increasing, well-founded demands to improve the depth and breadth of food and nutrition information that is collected in the survey. These demands will likely compete with demands for other kinds of health-related information in NHANES. We were not charged to assess those competing demands, but we offer a range of approaches to meet the needs for improved food and nutrition data from NHANES, some of which do not require additional data collection from NHANES respondents directly (for our recommendations, see Chapter 5).

We begin by describing efforts that the Economic Research Service (ERS) already has under way to enhance the information on diet and nutrition in NHANES and then discuss five other possible enhancements: special supplements for subsamples, links to food assistance program records, links to geographic information on food outlets, links to neighborhood characteristics, and links to price information. Some of these enhancements pose concerns of confidentiality protection and data access, which we briefly address at the end of the section.

ERS Supplement Initiative

Questions on household food expenditures and shopping patterns from the CSFII were not part of the original integration of the CSFII with NHANES, nor did that integration include the questions on diet and health knowledge and behavior that were part of the Diet and Health Knowledge Survey supplement to the CSFII. Yet the DHKS questions clearly provided useful information for policy making. Moreover, although the food expenditure and shopping data collected in the CSFII were not very detailed, they provided useful information on where people shopped, where they ate away from home, and how much they generally spent. Such data could be used to understand food stamp (or WIC) purchases and, more broadly, purchases by people who receive public assistance and those who do not. CSFII-type food expenditure data, together with the integrated NHANES dietary intake data, could also be used to understand the relationship between total household spending on food and what each person in the household consumes. Moreover, the health data in NHANES would make it possible to conduct analyses that trace through shopping patterns to food intakes to health outcomes. Participants in a workshop on the integrated NHANES-CSFII called for including the DHKS questions and other improvements in the data related to food and nutrition (Dwyer et al., 2003).

In response to the needs that were previously met by content that was part of the CSFII and the DHKS but not integrated with NHANES, the ERS is working with the National Center for Health Statistics to develop a

Flexible Consumer Behavior Survey Module (FCBSM). The full module will be included in NHANES in 2007 and 2008 and a scaled-down version in 2005 and 2006. The FCBSM will include questions on food shopping, food expenditures, self-assessment of diet quality, frequency of eating food away from home, attitudes toward and knowledge about diet and food safety, use of food labels, and safety-related preparation practices (e.g., frequency of washing raw foods). The module will take questions from the DHKS and the CSFII; it is also likely to include new or revised questions and questions drawn from other surveys. The intent is for the FCBSM to be a continuing supplement with a set of core questions and some questions that might change from year to year to meet emerging data needs.[3]

The work on the FCBSM is a positive development toward addressing gaps in data coverage for food and nutrition policy analysis and planning. A careful, thorough research and development program for developing content for the core FCBSM will be important so that the module provides information of most value at least burden on respondents (see Chapter 5).

Special Supplements

In addition to, or as part of, the Flexible Consumer Behavior Survey Module, supplementary survey questions could be given to subsamples of the full NHANES sample. This approach would be a means to obtain more detailed information on selected topics while not increasing the respondent burden. For example, a subsample could be asked to provide more detailed information on income, assets, food expenditures, food purchasing practices, and participation in food assistance programs.

A problem with a subsampling approach is that NHANES is already small for some subgroup analyses (such as subgroups of adults, adolescents, and children) even when pooling 2 or more years of data. One possible approach would be to rotate supplementary topics across years for the entire sample so that, for example, detailed questions on assets might be included in one 2-year cycle and detailed questions on income and program benefits in a subsequent 2-year cycle.

---

[3]Personal communication from James Blaylock, Economic Research Service, USDA, June 22, 2005.

Links to Food Assistance Program Records

An important set of policy concerns for USDA involves the costs, coverage, and effects on poverty, health, and nutrition of its food assistance and education programs. In recognition of the extensive data needs for analyzing programs, the Food and Nutrition Service and Economic Research Service of USDA have long supported targeted surveys and experiments involving program participants and other low-income households. These data have been used to study many aspects of the Food Stamp Program and other major food assistance programs, such as WIC and school feeding programs (see Hamilton and Rossi, 2002; Logan, Fox, and Lin, 2002; Fox, Hamilton, and Lin, 2004a, 2004b).

NHANES, for which the primary focus is monitoring health conditions for the general population, will never be that good a vehicle for analyses that compare participants with eligible nonparticipants in food assistance programs. Moreover, NHANES does not collect sufficient information on income, assets, and expenditures with which to estimate precisely who is and is not eligible for food assistance among the low-income population.

Yet NHANES does include data on participation in major food assistance programs, which can provide independent variables to include in econometric analyses to understand factors that relate to better and worse conditions of health and nutrition. However, except for last month's food stamp benefit, there are no NHANES data on benefit amounts or patterns of receipt, such as whether households participate in more than one program at the same time or different times, spells of participation, or who in the household is covered. Moreover, surveys tend to underestimate program participation by substantial amounts. For example, Cody and Tuttle (2000:21) found underestimates of participation in the Food Stamp Program of 26-37 percent in the March Current Population Survey Income Supplement for 1989-1999. Taeuber et al. (2004) found large differences between reporting of food stamp receipt in the 2001 Supplementary Survey (a predecessor to the new American Community Survey) and in matched records for the state of Maryland, due largely to underreporting by household respondents.

A low-cost means to obtain more detailed, accurate information on food assistance program participation in NHANES could be to match administrative records to the NHANES sample and append relevant program variables to the NHANES household and person records. An ERS-sponsored Survey of Food Assistance Information Systems that was conducted by mail in 2002 of program directors in 26 states found that states maintain Food Stamp Program and WIC data systems that are generally updated in real time, but they do not maintain such systems for school feeding programs (Cole, 2003; see also Cole and Lee, 2004). There is considerable experience in linking Food Stamp Program data with other record systems and surveys; linkages of WIC data have rarely been conducted.

Although NHANES collects Social Security numbers from sample members, consent has never been sought to use the numbers for record linkage. Until such consent is sought, linking Food Stamp Program records to NHANES records would require probabilistic rather than exact matching, by using such variables as name, date of birth, race, ethnicity, and, possibly, household income. Probabilistic matching would be required for WIC in any case, because, unlike the Food Stamp Program, most WIC programs do not require Social Security numbers from participants. Such matching, using software from the U.S. Census Bureau, was successfully performed in three states for a study of joint participation in food stamps and WIC (Cole and Lee, 2004).

Many analyses of food assistance program effects, whether with linked NHANES data or other sources, make use of empirical strategies that rely on instrumental variables to deal with possible bias—variables that are correlated to a potentially endogenous variable (for example, program intake), but are not themselves associated with the outcome of interest (for example, the nutritional health of food assistance program recipients). For example, if unusually effective mothers participate in WIC, then estimates of WIC program effects are likely biased upward; the opposite bias would exist if unusually ineffective mothers participate in WIC. One way to determine the extent and direction of this bias would be to try to use variation in state- and local-level administrative practices in WIC offices. If these practices occur roughly randomly, then variation in administrative practices could be used as instrumental variables for WIC participation. Some information is collected on these practices and has been used in some applications. With greater attention to collecting these types of administrative data, systematic research to enhance understanding of the behavioral effects of WIC might be feasible.

More generally, it would be worthwhile to devote systematic effort to obtaining geographic detail on administrative practices that vary across locations to use for analysis of food assistance program effects on diet and health.

Links to Geographic Databases on Food Outlets

A question of policy interest has been the availability of food shopping outlets that provide a range of reasonably priced, healthy food choices in comparison with the availability and concentration of fast-food and convenience store outlets that may not provide healthy alternatives. The 1996 National Food Stamp Program Survey obtained responses from program participants and other low-income households on the location of stores where they usually shopped and the supermarkets nearest their homes. These addresses were geocoded to latitude and longitudinal coordinates and the resulting information used to calculate distances to the nearest supermarkets and food outlets actually used. The results indicated that most low-income households use supermarkets as their main type of food store and do not typically face barriers to shopping at supermarkets (Ohls et al., 1999:xiii-xiv).

ERS proposes to include questions about food shopping habits in the new Flexible Consumer Behavior Survey Module to be added to NHANES. It would be burdensome for the FCBSM to obtain the level of detail in the 1996 survey on locations of specific food outlets used by households. However, with the advances in content availability on the Internet, a possible approach would be to link NHANES records to geographically based information on eating establishments and food retail outlets. This approach would require that the NHANES household addresses be geocoded by using the U.S. Census Bureau's TIGER (Topologically Integrated Geographic Encoding and Referencing) System or a commercial system based on TIGER. The same geocoding would then need to be done for the addresses of retail food outlets of various types, as well as fast-food establishments, sit-down restaurants, and other away-from-home options in cities or counties in the NHANES sample, using on-line directories and maps. Information might also be added from directories on the price range for eating establishments. The addition and regular updating of geographically based information on food outlets and eating establishments to NHANES records could provide valuable—if admittedly crudely measured—input for analyses of the environmental context of food decision making and changes in that context over time.

Links to Neighborhood Characteristics

In addition to adding information about nearby food shopping and eating establishments for households in the NHANES sample, it would be useful to add other characteristics of sample households' neighborhoods to permit contextual analyses of various kinds. Historically, the decennial census long-form sample has provided information on demographic, social, and economic characteristics for counties, cities, and neighborhoods (census tracts and groups of blocks). The 2000 census included a long-form sample of about one-sixth of all households (16 million records), but planning for the 2010 and future censuses assumes that the census will use only a short form with basic demographic information (age, sex, race, ethnicity, household relationship, and housing tenure). The new American Community Survey (ACS), which was fully implemented in early 2005, includes the content of the former census long form. The ACS questionnaire is sent monthly to a sample of 250,000 households, for an annual sample of about 3 million households. To provide estimates of sufficient reliability for areas with fewer than 20,000 people, 5 years' worth of data will be accumulated and averaged.

It would be useful to begin planning now on how to incorporate ACS estimates of neighborhood characteristics into the NHANES database. Such neighborhood characteristics as poverty ratios, educational levels, distributions of housing and utility costs, and ethnic composition would be useful contextual variables to include in analyses of food expenditure and consumption patterns. Because neighborhood-level data from the ACS will not be available until 2011 (for the period 2006-2010), it would be useful to consider incorporating some 2000 census neighborhood characteristics into NHANES.

Links to Price Information

Ideally, NHANES would provide information not only on food expenditures, but also on prices for specific types of foods for households included in the NHANES sample. Yet it would be very difficult and burdensome to collect information in NHANES about the prices that households pay for food. The aggregate data on food prices that are available through the Bureau of Labor Statistics (BLS) price index program could be used to augment household records in NHANES.

BLS publishes Consumer Price Indexes (CPIs) for food at home and food away from home by region (Northeast, Midwest, South, and West), population size classes of metropolitan statistical areas (MSAs), and the 26 largest MSAs. Data are collected, but not published separately, for 87 other geographic areas, not including rural areas; separate indexes for types of foods are provided only at the national level (see www.bls.gov/cpi/cpifaq.htm [June 2005]) . These indexes can properly be used only to compare rates of change in prices across areas—not price levels—because the data come from a probability sample of prices that is designed to produce the national CPI and so there is no particular consistency across areas in items that are priced.

BLS and the Bureau of Economic Analysis (BEA) have a research program to reanalyze the price data for geographic areas to develop fixed-weight interarea price indexes for major commodities that can be used to compare relative costs across areas (not just relative rates of change in prices). The approach uses hedonic regression methods to determine the contributions of geographic locations to the prices of various items (see Aten, 2005; Kokoski, Moulton, and Zieschang, 1999). ERS could investigate the geographic area pricing work by BLS and BEA to determine if appropriate relative price indexes for food at home and away from home for specific metropolitan areas are available that could be added to household records in the NHANES database. If so, the indexes could be used to study such issues as whether consumption of fruit and vegetables is related to areas with generally higher food prices.

The analytical uses of area food price indexes would be limited for several reasons. One is that they lack specificity by type of food, although it is possible that the BLS/BEA work could be disaggregated for some types of foods. Another is that their use would necessarily involve an assumption that food prices differ more among metropolitan areas than within them, which may not hold true in some or most areas. The results in Aten (2005) for food and beverages show a considerable range in price index values among areas—from 0.85 (Cincinnati) to 1.29 (New York City), with the mean value set at 1.00. This range (0.44) is higher than the price index value range for transportation (0.29), about the same as the price index value ranges for recreation and apparel, and lower than the price index value ranges for housing (0.79) and out-of-pocket medical care (1.38). While not addressing the issue of intra-area variability in food prices, for which data are not readily available, the BLS/BEA work suggests that the addition of area price indexes for food to NHANES records could be useful for some kinds of research. It would be a very low-cost enhancement to the data.

Another possible avenue to explore for adding food price information to NHANES involves sales outlet and household scanner data collected by private market research firms. If such data could be obtained at the city or neighborhood level and appended to NHANES records, the potential for understanding food consumption behavior would be greatly expanded. Scanner datasets have serious limitations, their quality is largely unknown, and making them usable for analysis purposes could be difficult. Nonetheless, scanner data are unparalleled in the detail they provide on prices for specific foods (see Chapter 3).

Confidentiality and Data Access

Many of the enhancements to NHANES that we outline in this section represent low-cost improvements for research on understanding household food expenditures and consumption by linking NHANES records with other data sources. Such data linkages, however, may increase the risks that individual households could be reidentified in public-use microdata files.

Because of the small sample size and clustered design for the continuing NHANES, in which only 15 primary sampling units are in each year's sample, special precautions are already taken to safeguard confidentiality: data are not released for single years but instead are pooled over 2 years, data masking steps are used on specific variables, and some variables (for example, age of nonsampled household members) are not included in any form on the public-use microdata files. Because of the 2-year pooling procedure, food and nutrition-related data that began to be collected in 2002 when the CSFII was integrated with NHANES are not yet available. Such items include the second day of dietary intake information and some items included for the first time in the first day of dietary intake, such as where each food was obtained; these items will first be publicly available in the 2003-2004 NHANES release.

The National Center for Health Statistics has two mechanisms for providing user access to data (such as geographic detail and day two dietary intake in 2002) that are not provided on public-use microdata files or that are altered in some way. Researchers may be approved to access those data at the NCHS research data center (RDC) at its headquarters in Hyattsville, Maryland, or remotely by e-mail, submitting SAS code to produce tables or regression coefficients, for example. NCHS staff review output from either the RDC or remote access submissions for confidentiality protection. There is a fee for using the RDC or the monitored remote access service; the fee for monitored remote access is reduced for users who plan repeated analyses of selected datasets that have been developed for frequent, multiple use (for the details of the NCHS policy, see www.cdc.gov/nchs/R&D/rdcfr.htm

[June 2005]). USDA could consider working with NCHS to develop low-cost access to NHANES records linked to other data at the NCHS RDC or by monitored remote access, perhaps developing special extracts oriented to food consumption analysis.

CONSUMER EXPENDITURE SURVEY

The Consumer Expenditure Survey (CE), which has been conducted on a continuing basis by BLS since 1980 to monitor the purchasing activities and habits of American consumers, is the most comprehensive source of expenditure data collected by the federal statistical system. (Predecessor surveys were conducted in 1972-1973, 1960-1961, and at intervals of about 10-15 years back to 1901; see Jacobs and Shipp, 1990.) CE data are used to estimate market basket weights of goods purchased for the Consumer Price Index. The CE is comprised of two surveys: (1) a diary survey, in which households keep two weekly expenditure diaries; and (2) a quarterly household interview survey, which obtains information for households on major purchases on a quarterly basis over the course of a year together with demographic, employment, and income information for the household members (Bureau of Labor Statistics, 2004b).

The diary survey is a record of daily expenses for a consumer unit (members of a household that share living expenses) that is kept by a respondent from each consumer unit for two consecutive 7-day periods. About 7,500 consumer units are surveyed each year. All daily expenses, except business expenses and expenses incurred while out of the home overnight, are included in the diary. The diary survey also collects demographic, work experience, and earnings data on household members aged 14 and over together with household income by source in the last 12 months.

Food-related items in the diary survey include usual weekly expenditures at supermarkets and specialty food stores together with detailed information on food purchases in each 1-week diary. The diary provides space for food purchases away from home by type of outlet and meal (for example, breakfast or lunch from a fast food outlet, vending machine, or full-service restaurant) and food purchases for consumption at home by type of food (for example, grain products, beef, sugars, vegetables—see Table 2-1). For each purchase, the respondent is asked to record the item, its cost, and some information about the form of the item (for example, canned, frozen, fresh).

The quarterly household interview survey collects information from about 7,500 consumer units every quarter for five consecutive quarters on expenses over the 3 months prior to the survey. It is designed to capture information about major purchases—including vehicles, major appliances, housing costs, and vacation costs—that are not generally available through the diary survey. This survey also collects information on housing characteristics, household appliances, ownership of real estate, work experience of household members, sources and amounts of household income, and information on financial assets, such as savings accounts, stocks, bonds, and mutual funds.

The household interview survey includes questions to double-check food purchases reported in the diary portion of the survey, including questions about food purchased away from home. Data from the two surveys are integrated to provide information about detailed day-to-day purchases, as well as long-term, major purchases.

Uses and Limitations

CE data have been used for analyses related to food and nutrition: for example, expenditures on fruits and vegetables by low-income households (Blisard, Stewart, and Jolliffe, 2004); household expenditures on vitamins and minerals (Lino et al., 1999); trends in food purchases away from home (Paulin, 1995); and factors related to food expenditures for use in projections (Blisard, Variyam, and Cromartie, 2003). The primary advantage of the CE is its rich data on all types of expenditures; in addition, it obtains relatively rich data on household employment and income.

However, the CE also has disadvantages for food and nutrition-related analysis. With regard to food consumption, the CE does not obtain information on actual dietary intake. Moreover, the food expenditure information in the CE does not always well describe the foods purchased or their quantity, so that the CE cannot be used as a source for prices paid for specific amounts of individual food items. For example, a survey respondent may report purchasing milk, but may not report what kind of milk—whole, low fat, or skim. The respondent may at the same time report purchasing milk for $2.19, but information on how much milk was purchased is not systematically collected. The CE also does not obtain information on who in the household consumed the food or how it was prepared. The CE's relatively small sample size further limits its potential for analysis. The CE data on expenditures and income also exhibit underreporting problems: in comparison with BEA's personal consumption expenditures series (PCE), the CE underreports spending in many expenditure categories. In particular, the aggregate CE amount for food for 1992-2000 averaged 72 percent of the corresponding PCE aggregate (see Garner et al., 2003). Income is also underreported, particularly for low-income families for whom reported expenditures often exceed reported income (Meyer and Sullivan, 2004).

Possible Improvements

Enhancements to the Consumer Expenditure Survey similar to those we suggest above for the integrated NHANES could extend the usefulness of the CE data, particularly the diary survey, for food and nutrition-related research and policy analysis. For example, reporting of Food Stamp Program participation could be validated and enhanced by matching CE records with the program's administrative records. Neighborhood characteristics from the 2000 census and the American Community Survey could be added to the CE records, as could links to geographically based information on retail food outlets. Although the CE is one of the most burdensome federal surveys, it might be possible to occasionally include supplemental questions related to food purchasing and consumption behavior that would enhance the value of the data for food and nutrition-related research.

To the extent that these initiatives increase the possible risks of disclosing confidential information, researchers would need to access the data at the BLS research data center at its headquarters in Washington, DC. It might also be possible to arrange for researchers to access enhanced CE microdata files on their own computers through a licensing agreement (see National Research Council, 2005a).

3

Proprietary Data Sources

A number of rich datasets on food purchases and consumption are produced outside the federal government's data collection efforts. These datasets comprise information collected by private market research firms in order to analyze food and related markets. Two types of data are typically collected: scanner data, which record sales of food purchased at stores or food used by consumers in their homes; and surveys of households that collect information on what is consumed by the household, either through direct questions about food consumption or through food diaries (see Table 1-1 in Chapter 1). These datasets contain an enormous amount of information about food purchases and consumption, including prices paid by consumers for food. Because the analyses conducted using these datasets are for firms interested in understanding the latest market trends, the data are usually available within a couple of weeks of collection. This chapter briefly describes some of these datasets and their key attributes, potential uses, and limitations for food and nutrition policy planning and research by the U.S. Department of Agriculture (USDA) and other agencies.

SCANNER DATA

Scanner data come from two types of data collections: (1) point-of-sale (retail) collections, which use the universal product code (UPC) of products sold at retail checkout counters to identify products and quantities sold and their prices; and (2) household scanner panels, which are usually random samples of households in which household members are asked to scan in the UPC of the items they have purchased, using scanners provided to them (see Box 3-1 for a summary of the data content of household scanner panels).

ACNielsen (formerly, A.C. Nielsen Company) and Information Resources, Inc. (IRI), are the two major producers of these types of datasets. For point-of-sale data, ACNielsen and IRI purchase price and item data from the scanner systems of cooperating retail outlets (the ACNielsen data collection is called Scantrack Services; the IRI collection is called Custom Store Tracking). Supermarket scanner data do not include fruits and vegetables, some prepared foods, and other products that lack UPC codes. They also do not cover restaurants or other food outlets.

Household scanner panel data are generated by randomly selected households, in which a household member scans in the household's food purchases from all types of stores over a week's time. As currently designed, these data provide limited demographic characteristics. Information collected on products with a UPC includes price, quantity, and promotional information. For items that lack a UPC, such as meat and fresh produce, participants are asked to identify the type of item and its weight. Both ACNielsen and IRI conduct these types of panel surveys for nationally representative samples of more than 61,500 and 50,000 households, respectively (the ACNielsen data collection is called the HOMESCAN Consumer Panel; the IRI collection is called the Combined Outlet Consumer Panel).[1]

Researchers inside or outside the government must purchase scanner data, although the cost need not be high, depending on the amount of data required. A study by the Food and Nutrition Service estimated the cost of 2 months of scanner data collection for a supermarket chain to be $35,000. To consider this figure in context, the study indicated that the National Survey of Food Stamp Program Participants in 1996 cost $1.7 million. Purchase of the necessary scanner data for applications that required many months or years of observations for many outlets could, of course, entail substantial costs.

---

[1] Only one-quarter of households in the HOMESCAN Consumer Panel are asked to record items that lack a UPC code.

PROPRIETARY DATA SOURCES 55

BOX 3-1
Household Panel Scanner Data

What?	Who?	Where?
• Price Paid • Quantity Purchased • Purchase Date • Product Category • Brand • Size • Universal Product Code (UPC) • UPC Description • Coupon Information • Product Attributes — Flavor — Form — Fat Content — Sodium — Cholesterol — Organic — Container Type • Store Name Identifier • Channel Type Identifier	• Household Size • Household Income • Female and Male Head Characteristics — Age — Education — Employment — Occupation — Marital Status — Race • Household Composition • Presence and Age of Children • Local Market Identifier • Region • Projection Factor (Weight)	• Grocery Store — Kroger, Safeway, etc. • Drugstore — CVS, Walgreens, etc. • Mass Merchandiser — Target, Value City, Wal-Mart, etc. • Supercenter — Big K, Super Target, Wal-Mart Supercenter • Warehouse Club — Costco, Sam's Club, etc. • Convenience, Gas • Other — Dollar Store — Farmers' Market — Online Purchase — Etc.

Uses

Scanner data have been used in published economic studies for over a decade to answer a variety of questions about food consumption, food pricing, and the operation of retail food markets. Most applications to date have used retail data; a few have used household data or a combination of the two. Scanner data have been used most often to examine pricing behavior in particular product markets, including the influence of private-label foods on name-brand pricing (Putsis and Cotterill, 2001; Ward et al., 2002), strategic pricing responses in markets supplied by only one or two firms (Vickner and Davies, 2002), and the effect of political pressure on breakfast cereal prices (Cotterill and Franklin, 1999). Scanner data have also been used to measure the value of product attributes (Bonnet and Simioni, 2001; Unnevehr and Gouzou 1998), assess bias in the construction of the Consumer Price Index (Reinsdorf, 1999), analyze seasonality in prices and consumption (Chevalier, Kashyap, and Rossi, 2004; MacDonald, 2000; Thompson and Wilson, 1997), and develop basic estimates of price elasticities for specific food products (Jones, 1997; Maynard and Veeramani, 2003). In studies to estimate price elasticities, income is controlled imperfectly through store location. Scanner data have been used for policy-relevant food and nutrition research, such as studying the effects of mandatory nutrition labeling (Mathios, 1998, 2000) and the redemption activity of food stamp and cash assistance clients in conjunction with the Maryland demonstration project on electronic benefit transfer (Cole, 1997).

Finally, scanner data have been used for general descriptive work to answer such questions as, for example, whether fresh fruits and vegetables are more expensive than processed fruits and vegetables and how much it costs to meet guidelines for daily intake of fruits and vegetables (Reed, Frazao, and Itskowitz, 2003). Thus, scanner data have the potential to address questions related to market sales, price response in markets for very specific products, how pricing relates to product characteristics including specific nutrition characteristics, firm behavior in concentrated processed food product markets, and consumer demand for specific kinds of food products.[2]

---

[2] A workshop on the uses of scanner data in policy analysis, organized by the Economic Research Service, USDA, and the Farm Foundation, included useful reviews of the advantages and limitations of scanner data (www.farmfoundation.org/projects/documents/ScannerDataWorkshopSummaries2_000.pdf [June 2005]); see also Appendix A in this report.

Scanner datasets contain several valuable attributes that make them attractive for some specific uses. One of those key attributes is detailed information on the product that was purchased, including the brand name of the product, the exact description of the product (for example, for orange juice, whether it is calcium enriched), the quantity of the product, and the price for which the product was purchased, including whether it was on sale or part of a promotion. This linkage of price and detailed quantity and product data for individual household purchases is unique among all the datasets reviewed in this report. Another key attribute of these data is that they are produced in a timely manner, unlike those from federal surveys. Furthermore, the household scanner panel samples are much larger than those for the Consumer Expenditure Survey (CE).

Limitations

Scanner data do have limitations. Coverage is a key issue for point-of-sale scanner datasets. Although most major retailers, including warehouse clubs (Sam's Club, B.J.'s, Costco), participate in both the ACNielsen and IRI point-of-sale collections, the largest retailer in the nation, Wal-Mart, does not. Some smaller mom-and-pop grocery stores do not participate, either. In addition, as noted above, some items do not have UPC codes, including fresh vegetables and fruits, meats, baked goods, and other prepared foods.

The household scanner panels cover only food and other items purchased in retail stores, not food purchased in restaurants. Moreover, many households in a given week will not have purchased specific products, raising problems for how analyses should deal with infrequent purchases and the frequency with which people shop. Studies that have linked retail and household scanner information have encountered inconsistent data between the two data sources.

In addition, the household scanner surveys place a big burden on respondents. A respondent is asked to scan in all the items purchased after each shopping occasion and report the results to the collecting firm. Households are sent scanners with guidelines or training videos on how to use them. Unlike the CE survey, which is an in-person interview, interviewers do not go through the data collection procedure with the household members, although a telephone helpline is available.

Households in the IRI Combined Outlet Consumer Panel are asked to scan all their purchases from stores. They receive points that can be exchanged for prizes, vacations, and in restaurants for every shopping trip for which purchases are scanned, and they can participate in the survey as long as they wish. Participants in ACNielsen's HOMESCAN panel are asked to transmit data on scanned purchases through a regular telephone line once a week. HOMESCAN panelists also receive points that can be redeemed for prizes for each data transmittal (personal communication, G. Crusafulli, ACNielsen). Typical response rates for the HOMESCAN panel are around 85 percent. IRI does not publicly release response rates for its household scanner panel survey.

Representatives from ACNielsen presented information about the HOMESCAN survey at the panel's workshop (see Appendix A). They reported that they have trouble recruiting some groups to participate in the survey, specifically, young single adults, people in low-income households, and minorities. Jensen (2003) compared a sample from the HOMESCAN panel to U.S. national averages from the 2000 census and found that the HOMESCAN sample households by comparison had higher incomes, were smaller, were more likely to be married couples, were more likely to be white, and were less likely to be Hispanic (see also Appendix A in this report). Thus, these data may not be useful for specific analyses of underrepresented groups. The IRI Combined Outlet Consumer Panels also tend to overrepresent higher income households in comparison with 2000 census data.

The household scanner panels are not designed to collect much information on the households selected for the sample. Some basic demographic information is collected, but it is not very detailed. No information is collected on health, physical activity, or diet and health knowledge. Although data on employment status, total household income, and vehicle ownership are collected, information about assets, sources of income, and participation in food assistance programs is not collected.

One general limitation of point-of-sale scanner data and household scanner data is that the UPCs do not always clearly identify items. The codes are 10-digit numbers that are intended to be universal guides to products sold. The first five digits for an item are assigned by the centralized Uniform Code Council, and the last five digits are assigned by the corporations that make the product. Guidelines are given to corporations to help them assign the last five digits, but there is evidence that these guidelines are not necessarily followed and that codes change for some products. This phenomenon has implications for the difficulty and accuracy of placing products into specific categories. For example, Mladenic, Eddy, and Ziolko (2001) in an analysis of more than 280,000 UPCs for grocery products found that 44 different codes were used for fresh 2 percent milk. Thus, any effort to identify specific products consumed would need to work through potentially difficult coding issues.

Future Potential

Despite the limitations of the retail and household panel scanner datasets, many researchers, both in USDA and in academic and private research organizations, have begun to exploit scanner data because they provide such extensive detail on food products, quantities purchased, and prices. None of the food and nutrition-related datasets produced by the federal government can match the scanner datasets on this type of content, nor on the timeliness with which they are produced. Because of these advantages, it is hard to exclude them as potential sources of information for USDA policy and decision making.

Before placing significantly greater reliance on scanner data, however, additional work must be done to examine the characteristics and representativeness of the population covered by the data and other possible sources of bias (see Kirlin and Cole, 1999). If the research on data quality supports the usefulness of scanner data, they could be drawn on to examine a wide range of issues.

Specifically, the retail scanner datasets could be used to examine short-run and long-run aggregate market trends. They could also be used to compare aggregate totals on food purchases with other sources of data on food expenditures—for example, from the CE survey and from the national food disappearance data, which measures the flow of raw and semiprocessed foods through the U.S. marketing system.[3] In addition, the retail scanner datasets could be linked with data on club card members to obtain some very basic information on the households that purchase the goods.

The household panel scanner datasets could be used to understand short-run and long-run trends in foods consumed by households and the relationship between price and consumption. The level of detail on products purchased could allow for analysis of consumption trends when new

---

[3]The USDA "national disappearance" estimates, produced annually, provide estimates with a 2-year lag of commodities that are *available* for food purchase and consumption. They are developed on the basis of production estimates adjusted for inventory changes, exports, imports, and nonfood uses (see www.ers.usda.gov/Data/FoodConsumption/FoodAvailDoc.htm [June 2005]).

products are introduced or when minor changes to products are made. The household scanner data could also be used to understand aggregate changes in purchases in relation to changes in policies on healthy eating, such as changes in the food pyramid guidelines that were announced by the USDA in January 2005, or to food safety recalls.

If information on participation in food assistance programs (for example, food stamps, WIC, school breakfast and lunch) could be added to household scanner data, the augmented datasets could be used to track and compare expenditures of food assistance program recipients and of nonrecipients with similar incomes. With augmented household scanner data it might also be possible to address such questions as why the participation rates among the eligible population in the Food Stamp Program plummeted in the 1990s. Was this phenomenon a by-product of the expanding economy and welfare reform, or was it due to changes in food preparation and consumption behavior or both? Specifically, with the rise of labor force participation rates among women (both single and married) over the past decade, the time that is available to prepare foods for home consumption has declined, and major grocery stores have significantly expanded the quality and quantity of prepared food items. However, one cannot use food stamps to purchase prepared foods. Is part of the low rate of food stamp use a by-product of the fact that families have less time to prepare foods and that grocery stores provide attractive alternatives not available to food stamp recipients?

HOUSEHOLD FOOD CONSUMPTION SURVEYS

Two other major food consumption surveys are conducted by the NPD Group, a sales and marketing research firm. The National Eating Trends (NET) Survey obtains food intake data from a nationally representative sample, and the Consumer Report on Eating Share Trends (CREST) collects information from a large online sample of consumers on their purchases of prepared meals and snacks at commercial restaurants and other outlets. Both of these datasets are used in analyses by firms interested in food market trends.

The NET survey has been conducted since March 1980. Over the course of a year, 2,000 households record diaries of food and beverage consumption for 14 consecutive days for all individuals in the household. The survey questionnaire and diary are mailed to 60 new households every

Monday. Data are usually processed and available for analysis within three months of collection.

In addition to the food intake diaries, the NET survey collects information on the types and brands of foods consumed, how they were prepared and served, the ingredients used in home-prepared meals, and who in the household consumed them. Information is obtained on whether the respondents were on a diet during the 14-day period and which type of diet they were on, whether they have any medical conditions, their height and weight, supplement use, exercise level, and attitudes on nutrition. Some demographic information is also obtained on respondents.

The CREST survey is an online survey that collects information from 3,000 adults and 500 teenagers on a daily basis (42,000 responses per month). Survey respondents are asked to report what they ate, where they purchased it, where they ate it, who they were with, and how much they spent for food at commercial outlets the day before the survey. The survey also includes behavioral and attitudinal questions.

These surveys collect unique information that could be useful in a number of policy environments. The CREST survey's unique focus on food eaten away from home could fill in gaps from other surveys on what is known about eating out. However, since the survey is an online survey, it will not cover those without Internet access. Thus, these data may not be useful for low-income or elderly populations. The survey also has low response; typically the response rate is just over 40 percent.

The NET data are unique in providing 14 days of dietary recall, which is an extraordinary amount of information on food intake that is not matched in any other dataset. This information could be used to provide more stable estimates of consumption of different types of food than the two-day recalls from NHANES. It might also be useful for estimating consumption of foods that are eaten less regularly, which may be critical for certain food safety risk assessments. Information about preparation of food and ingredients used could also be used in food safety risk assessments. The other key attribute of these data is that they include information on attitudes towards food and dieting practices. This information, if released in a timely manner, could be useful in picking up market trends related to dieting practices. For example, the recent popularity of the Atkins and related diets is believed to have had large effects on major food purchases, such as meat, grains, and fruits. Timely information about dieting practices might be useful for analyses of these trends.

Of greatest concern with the NET data is the quality of data collected through the 14 days of dietary recalls. Since this amount of recall places significant time and recall burdens on respondents, the quality of the data may suffer. This issue would need careful scrutiny before basing important public policy decisions on results from NET-based analyses.

4

Other Federal Data Sources

A number of other federal datasets contain information that is relevant to food consumption and nutrition monitoring and could be used to address policy issues. Most of these datasets have purposes that are not directly related to food and diet monitoring, but they do contain some useful information. In this chapter, we briefly review several of these data sources and discuss possible ways they could be used to address food policy questions. To add value for this purpose, consideration could be given to enhancing one or more of these datasets in ways similar to those suggested in Chapter 2 for the NHANES and CE surveys—for example, by matching survey records with administrative records for food assistance programs to add to or improve the quality of program participation data, by appending neighborhood characteristics of various kinds from the 2000 census, the American Community Survey (when small-area data become available), and various geographic databases, or by adding supplemental modules with additional food and nutrition-related questions.

Our discussion is organized around five types of data. They are: data on monetary resources for food consumption, including food insecurity (the Current Population Survey); time use data (the American Time Use Survey); data sources for longitudinal analysis of food consumption and related behavior over a span of years (the Early Childhood Longitudinal Study, the Health and Retirement Study, and the Panel Study of Income Dynamics); data sources for relatively quick-turnaround studies of emerging issues, which can also provide key trends for the nation, regions, and states (the Behavioral Risk Factor Surveillance System and the State and Local Area Integrated Telephone Survey); and a possible data source for analyzing food consumption behavior of the low-income population (the Expanded Food and Nutrition Education Program).

CURRENT POPULATION SURVEY

The CPS is an ongoing monthly survey of about 56,000 households, which is fielded by the U.S. Census Bureau and supported by the Bureau of Labor Statistics. (The Census Bureau and other federal agencies pay for periodic supplements to the main labor force survey.) Its primary purpose is to provide estimates of employment, unemployment, and other characteristics of the labor force. Each year, beginning in 1995, USDA has supported a supplement (currently fielded in December) on food expenditures, food assistance program participation, food insecurity, and ways of coping with not having enough food—see Box 4-1.

The CPS sample design is state representative, and its large size will support state-level estimates when the data are averaged over 3 years. Response rates are high, averaging 92 percent for the main labor force survey, although 12 percent of households do not complete the food insecurity module.[1]

The food expenditure, program participation, and food insecurity data from the December supplement can be analyzed with information from the main CPS questionnaire, which includes demographic characteristics for all household members, detailed information on labor force participation and usual hours worked and earnings for household members aged 15 and older, and total household income. Because of the rotating design of the CPS and the recent expansion of the March income supplement to include households in February and April, about one-half of the households in the December sample can be matched with the same households in February or March that have detailed income, program participation, and health insurance coverage information for the preceding calendar year from the renamed Annual Social and Economic Supplement.[2] The combi-

---

[1] Personal communication from Mark Nord, Economic Research Service, USDA, May 31, 2005.

[2] More precisely, the same addresses can be matched. If a household has moved, the match will not represent the same people (see www.bls.census.gov/cps/asec/2003/sdataqua.htm [June 2005]).

BOX 4-1
Food and Nutrition-Related Data in the December Current Population Survey

Where Bought Food Last Week
- Supermarket or grocery store
- Other places where people buy food (meat markets, produce stands, bakeries, warehouse clubs, convenience stores)
- Restaurant, fast-food place, cafeteria, vending machine
- Any other place

How Much Spent on Food Last Week
- Supermarkets and grocery stores (how much of total for non-food items)
- Stores such as meat markets, produce stands, etc. (how much of total for non-food items)
- Restaurants, fast-food places, cafeterias, vending machines
- Other places

Whether Would Need to Spend More or Less for Just Enough Food and How Much More or Less

Food Assistance Program Participation
- Receive food stamps in last 12 months and which months
- Amount of most recent food stamp benefit
- Children aged 5-18 receive free or reduced-price lunches at school in last month
- Children receive free or reduced-price breakfasts at school in last month
- Children receive free or reduced-price food at day care or Head Start center in last month
- How many women and children receive WIC foods in last month

Food Insecurity Scale Questions for Households with and without Children

Ways of Coping with Not Having Enough Food
- Receive meals from "Meals on Wheels" or other community programs in last month
- Eat prepared meals at a community program or senior center in last month
- Get emergency food from a food pantry, church, or food bank in last 12 months and how often
- Have a source of emergency food nearby
- Eat meals at a soup kitchen in last 12 months and how often

SOURCE: National Research Council (2005b: Appendix A).

nation of detailed income and program participation information and the food expenditure and insecurity data for matched households could support in-depth analysis of the effects of income constraints on food purchasing, the role of food assistance programs in alleviating food insecurity, whether people who lack health insurance coverage are more or less food insecure than households with public or private coverage, and similar topics.

AMERICAN TIME USE SURVEY

The American Time Use Survey (ATUS) is a relatively new, ongoing survey conducted by the U.S. Census Bureau for the Bureau of Labor Statistics (BLS) to measure how Americans spend their time. Data collection for ATUS began in January 2003 (see Abraham, 2004). Reports are produced quarterly and annually. The sample population is drawn each month from households in the outgoing rotation group of the CPS (see above), except that the sample is somewhat smaller, especially in less-populous states (those oversampled in the CPS). The ATUS household sample is stratified by race and ethnicity of the householder, presence and age of children, and number of adults in households without children. Households that have a Hispanic or black householder and households with children are oversampled. One respondent is selected randomly from those aged 15 and over in each household in the sample. The current sample size is about 26,000 households annually (see Herz, 2004; see also www.bls.gov/tus/home.htm [June 2005]).

Respondents are sent instructions and time diary materials and are assigned a day for which to report their activities. Interviews are conducted by telephone on the day following the assigned day. ATUS provides an incentive for households without telephones to call on the scheduled interview date, but the response rate for households without telephones is about one-half the rate for households with telephones, and the overall response rate is low: it was 57 percent in 2003 (Bureau of Labor Statistics, 2004a). Survey questions include review of the time diary; height and weight of the respondents, to permit calculation of the body mass index (BMI) and determination of being overweight or obese; identification of persons who were with the respondent during various times of the day; work summary questions; summary questions about secondary child care (that is, taking care of a child while doing something else); volunteering summary questions; and questions about trips away from home for 2 or more nights in a row. If respondents engaged in two or more activities simultaneously, they must identify which was the primary activity, and that activity is coded as the only activity (Herz, 2004). Because the sample comprises members of households that completed their eighth and final CPS main labor force survey, the employment and earnings data from that survey are available for use with the ATUS information. ATUS allows supplemental modules to be added. Modules are developed in cooperation with the BLS ATUS staff, and all modules' questions must be pretested to ensure that they elicit the desired information. Modules must run for a minimum of six months (Herz, 2004).

ATUS has many possible uses for food and nutrition-related research and policy analysis, and USDA's Economic Research Service (ERS) is currently working with BLS to plan a food and eating module to include in ATUS. The module is intended to provide data that can be used to analyze the relationship between patterns of time use and eating patterns, nutrition, and obesity, as well as food assistance program participation and grocery shopping and meal preparation.

One potential use for the proposed ATUS food and eating module is to estimate how much time it takes to prepare meals. Although time is not part of the "costs" calculated in USDA's four food plans (thrifty, low-cost, moderate, and liberal), data from the ATUS could be used to give a more robust estimate of the costs of meal preparation at various expenditure levels, including time costs. Another potential use of ATUS is to understand and improve access to food assistance programs by indicating how much time it takes to acquire, use, and retain program benefits, such as food stamps and WIC (Frost, 2004). Such data could also be used to understand how food consumption and preparation practices vary across households with different work schedules and arrangements. The time tradeoffs for meal preparation and the effects of recent policy initiatives for increased work effort among low-income populations could also be studied.

One feature of the food and eating module is that it will code eating as a secondary activity if, for example, an individual's primary activity is watching television. The module will also gather information on whether the individual was snacking on food (Hamrick, 2004). The ATUS food and eating module could be used to determine if snacking and the number of eating episodes throughout the day are increasing, whether it takes longer on average to prepare a meal at home than to eat away from home, what technologies and practices provide time savings for home meal preparation, and whether there are true time "shortages" that are leading to less cooking. There could also be questions that would provide information about the time it takes consumers to read and understand nutrition labels and ingredient lists.

Other possible uses are easy to imagine. They include how the number of eating episodes is related to obesity; how physical activity (or lack of it) affects eating patterns and obesity;[3] how income is related to the number of eating episodes, eating at home or away from home, and consuming previously prepared foods; how marital, social, and economic status relate to food shopping time; how food-related activities function as child care; demographic factors that relate to eating times; and which groups of people are able to eat during work and what effect (if any) this has on being overweight or obese (Hamermesh, 2004).

PANEL SURVEYS

For many kinds of analysis, particularly to inform policy planning, it is desirable to have measures on the same individuals over time. Longitudinal information from a panel survey that repeatedly interviews the same respondents would facilitate research on changes in food consumption behavior, diet, and health at the household or individual level and how they might relate to such factors as changes in income and program participation, initiatives for food education and safety, or changes in other contextual factors. Longitudinal data are usually expensive to collect, so that fielding a comprehensive new panel survey specifically for food and nutrition-related behavioral analysis does not seem feasible with the resources currently available to the Economic Research Service. However, several existing longitudinal surveys have the potential for use by ERS and other relevant agencies and perhaps could be enhanced to better support food and nutrition-related analyses.[4]

Early Childhood Longitudinal Study

The Early Childhood Longitudinal Study (ECLS) is sponsored by the National Center for Education Statistics in collaboration with several other

---

[3]See Dong, Block, and Mandel (2004) for an analysis of energy expenditure, using data from EPA's 1992-1994 National Human Activity Pattern Survey, which collected 24-hour time diaries by telephone.

[4]In addition to the three panel surveys described in the text, a number of other panel surveys could potentially be useful for nutrition-related analysis, such as the National Longitudinal Surveys of BLS (see Logan, Fox, and Lin, 2002).

agencies. This study follows two cohorts of children to collect information on young children and their family, school, and community environments. About 22,000 children at about 1,000 public and private, part-day and full-day kindergartens from the kindergarten class of 1998-1999 make up the kindergarten cohort (see West, Denton, and Reaney, 2000). Data were collected in the fall and spring of their kindergarten and first-grade school years and in the spring of their third- and fifth-grade school years, and it is planned to follow these children through twelfth grade. Questionnaires were administered to parents, teachers, and the children themselves (for details, see nces.ed.gov/ecls/kindergarten.asp [June 2005]). The birth cohort survey is following more than 10,600 children born in 2001. Data were collected through parent and caregiver questionnaires when the children were 9 months and 2 years old and will be collected when they are 4 years old and in kindergarten (for details, see nces.ed.gov/ecls/birth.asp [June 2005]).

The parent questionnaire for both cohorts includes the USDA's food insecurity scale questions, which provide information about the food security status of the children's families. Both cohort questionnaire sets include questions on the height, weight, and physical activity of the children and about participation in WIC and the Food Stamp Program. The parent questionnaire for the birth cohort at 2 years old asked about breastfeeding, formula use, and other beverage consumption by the child of interest. The parent questionnaire for the kindergarten cohort asks about school lunch and breakfast program participation, and the fifth-grade questionnaire asked children about their purchase of sweet and salty snacks and soda at school, and their consumption in the past 7 days of milk, juice, soda, carrots, green salad, potatoes, other vegetables, fruit, and meals at fast-food outlets.

Health and Retirement Study

The Health and Retirement Study is an ongoing panel survey of about 22,000 people who were aged 51 and over when they were first interviewed and their spouses. Blacks, Hispanics, and residents of the state of Florida are oversampled. The initial cohorts consisted of about 12,500 people aged 51-61 and about 8,000 people aged 70 and older in 1992-1993. New cohorts of people aged 56-61 and 69-75 and their spouses were introduced in 1998, filling in the entire older age span. The survey attained a steady state in 2004 with the introduction of another new cohort of people aged

51-56. New cohorts will be introduced in a similar manner every 6 years (sample sizes are smaller for newer cohorts than for the original HRS cohort).

Interviews are conducted in person the first year a household is included in the sample and by telephone every 2 years thereafter. Response rates for each cohort have been 80 percent or less at the original interview and 90 percent or more at subsequent interviews. The survey is conducted by the University of Michigan with funding from the National Institute on Aging (see hrsonline.isr.umich.edu [June 2005]).

The HRS collects comprehensive information on many characteristics: demographic background; disability; employment status and job history; family structure and transfers; self-reported health status and medical conditions; self-reported smoking, drinking, and exercise; cognitive status; health insurance and pension plans; housing; income and net worth; retirement plans and perspectives; and attitudes, preferences, expectations, and subjective probabilities related to retirement. The HRS also collects information on housing costs, out-of-pocket medical care expenditures, food expenditures per week or month, in stores and delivered, and expenditures for meals eaten out. Versions of the HRS are available under special access arrangements with links to Medicare and Social Security earnings and benefits.

In addition to questions that are asked at every interview, the HRS typically includes a large number of modules with supplemental or experimental questions, which could be used to ask questions related to food consumption and related topics. A special supplemental survey, the 2001 Consumption and Activities Mail Survey (CAMS), obtained relevant information on time use and spending. It was mailed to a random sample of 5,000 HRS respondents and there were 3,800 usable responses. CAMS covered time use (36 categories), spending (32 categories, including food-related items), and anticipated and actual changes in spending pre- and post-retirement (Hurd and Rohwedder, 2003).

The HRS is a source of extensive information about the financial well-being and health situation of older Americans and, more important, how their situations change as they age and experience such life events as retirement or loss of a spouse. Consideration could be given to enhancing the food and nutrition-related content of the core questionnaire beyond the limited information obtained on food expenditures, as well as to adding modules about food and nutrition. Such enhancements would support behavioral analyses of food consumption behavior and health effects for the older population.

Panel Study of Income Dynamics

The Panel Study of Income Dynamics (PSID) is a continuing panel survey of a cohort of families that began in 1968. The survey is sponsored and conducted by the University of Michigan Survey Research Center. Since 1983 the National Science Foundation has been the principal funder, with substantial continuing support from the Office of the Assistant Secretary for Planning and Evaluation in the U.S. Department of Health and Human Services and some support from other agencies, including USDA (see psidonline.isr.umich.edu [June 2005]).

The original PSID sample comprised two components: (1) 2,900 families drawn from the Survey Research Center national sampling frame, representative of the civilian, noninstitutionalized population; (2) 1,900 low-income families with heads under age 60 drawn from the 1966-1967 Survey of Economic Opportunity conducted by the U.S. Census Bureau. In 1990, 2,000 Hispanic families were added, but these families were subsequently dropped in 1996, and 441 immigrant families (including Asians) were added in 1997. Currently, more than 7,000 families (including original sample families and the subsequent families of their members) are interviewed once every other year, mostly by telephone (prior to 1997, interviews were conducted annually).

The PSID experienced a large sample loss—24 percent—at the initial interview in 1968, but additional sample loss dropped to 8 percent of the eligible families at the second interview, and it was only 1-2 percent at each interview thereafter. The extent to which attrition introduces bias into estimates from the PSID is not clear; some studies have reported little effect; others have found some biases in estimates of poverty rates before the new Hispanic sample was added in 1990 (see National Research Council, 1995:App. B).

The core content of the PSID includes many elements: family members' demographic characteristics; detailed employment histories and income by source for the household head and spouse; less detailed income information for other family members; program participation, including amounts and months received for food stamps; estimates of federal taxes paid; housing costs; average weekly food expenditures for home consumption and away from home; housework time; socioeconomic background; religion; and health status. Versions of the PSID are available under special access arrangements that contain geographic match codes for locations of PSID households down to the census tract level. With these match codes, researchers can append neighborhood data from the decennial census long-form sample or other files to the PSID records. The PSID has included supplements on many topics, including eligibility for food stamps and Supplemental Security Income, smoking and exercise, time use, and wealth, among others.

The PSID is the longest running nationwide panel survey in the United States with detailed socioeconomic information, and it has some information that is relevant for food and nutrition policy analysis. It could be worthwhile for USDA to explore ways to add questions to this rich data source.

QUICK-TURNAROUND SURVEYS

There may be some cases for which a few questions about dieting practices and attitudes or concerns about food safety will provide useful information for monitoring food market trends or food consumption behavior. For example, USDA may want to know how many Americans are practicing the Atkins diet or how people are reacting to stories about mad cow disease. Such information may be used to begin to understand a trend, particularly when policy makers do not want to wait for the results of other surveys that collect this information. Two data collection programs that are geared to the addition of questions and modules to track emerging trends—the Behavioral Risk Factor Surveillance System (BRFSS) and the State and Local Area Integrated Telephone Survey (SLAITS)—have the potential to serve this purpose. These programs also have the advantage that they can provide state-level estimates. A disadvantage is that they are limited to households with land-line telephones.

Behavioral Risk Factor Surveillance System

The BRFSS is an ongoing cross-sectional survey designed by the Centers for Disease Control and Prevention (CDC) and conducted by health departments of the states and territories with technical advice and oversight by the CDC (see www.cdc.gov/brfss/index.htm [June 2005]). The purpose of the survey is to track the health habits of the population. Information is collected by telephone interviews each month from a random sample of adults aged 18 and over in each state. CDC aggregates the results for each year and provides them to states to use to direct health promotion and disease prevention programs. The sample size is about 4,000 interviews per state.

The CDC initiated the BRFSS in 1984, at which time 15 states participated in monthly data collection. CDC developed standard core questionnaire for states to use to provide data that could be compared across states. By 1994, all states, the District of Columbia, and three territories were participating in BRFSS. The BRFSS was designed to collect state-level data, and a number of states from the outset stratified their samples to allow them to develop estimates for intrastate areas. The core survey, which includes fixed questions, rotating questions asked every other year, and several questions on "emerging issues," is reviewed every year and changes are made by CDC and the states working together.

All states administer the core survey and may also choose among several optional modules; they may also add their own questions. The core survey contains no food or nutrition-related questions except those regarding consumption of alcohol, a question on physical exercise, and self-reported height and weight. Several modules do have such questions. Optional Module 12, on cardiovascular disease, specifically asks if respondents are eating fewer high-fat or high-cholesterol foods and if they are eating more fruits and vegetables to reduce the risk of developing heart disease. Module 13, on folic acid, includes questions about supplement use and specifically whether supplements used contain folic acid. Module 19, on binge drinking, asks a number of follow-up questions to determine risky health behavior related to binge drinking.

The information currently available from Modules 12 and 13 can be used to understand reasons for changes in fat and cholesterol consumption, on one hand, and fruit and vegetable consumption on the other, as well as the use of supplements, multivitamins, and, specifically, folic acid. It could be possible to add questions to Module 13 if further information about supplement use is desired.

Other modules could be developed to address specific food and nutrition behaviors. Such modules could include a food frequency questionnaire designed to track food consumption behaviors with confirmed or suspected health risks or benefits, such as excessive consumption of seafood with high levels of mercury. The food frequency questionnaire could also provide information on overall eating practices that could be linked to the self-reported heights and weights in the core BRFSS that are used to calculate

BMI and the prevalence of obesity. Since the BRFSS is a sample survey within each state, however, modules can be added only through negotiations with each state.

State and Local Area Integrated Telephone Survey

Another possibility for obtaining food and nutrition-related information to track emerging trends would be to make use of SLAITS. SLAITS is a mechanism for government agencies and other sponsors to obtain customized state-level information by using the sampling frame from the National Immunization Survey (NIS). The NIS, in turn, is an ongoing telephone survey conducted by the National Center for Health Statistics (NCHS) that screens almost 1 million households per year to produce estimates of vaccination coverage levels among children aged 19-35 months.

NCHS will work with sponsors to design a specific questionnaire and sampling scheme that can be piggybacked on the NIS at any time. Typically, it takes 3-6 months of design and testing before data collection can begin. Sponsors can use previously developed SLAITS modules (which include health, child well-being and welfare, early childhood health, and asthma) or specify new modules. A SLAITS sample can be designed to target population groups, such as low-income households or those with specific characteristics, and NCHS will adjust the results for noncoverage of households without land-line telephones (see www.cdc.gov/nchs/slaits.htm [June 2005]).

EXPANDED FOOD AND NUTRITION EDUCATION PROGRAM

A dataset that could be used to understand food consumption in populations served by food assistance programs comes from the Expanded Food and Nutrition Education Program (EFNEP), which is part of USDA's Cooperative State Research, Education and Extension Service. The EFNEP program provides advice and counseling to low-income families with children through an experiential education program to improve nutrition, food resource management, and food safety behaviors. The goal is to enable participants to provide nutritionally adequate meals for themselves and their families. As part of the program, EFNEP participants fill out 24-hour food recall forms and a food practices checklist at initiation and at the completion of locally administered courses in which they learn such skills as food budgeting, selection, preparation, storage, and safety.

About 150,000 low-income families participate in EFNEP each year, and food records have been collected by EFNEP for about 100,000 individuals for each of the past 10 years. The food records are entered into an EFNEP Evaluation/Reporting System, which is used to determine overall diet quality, based on key indicators: total fat, protein, carbohydrate, fiber, calories, iron, calcium, and vitamins A, C, and B_6, as well as the number of servings of each of the food guide pyramid food groups. Also administered with the food recall is a 10-item food practice checklist covering other behaviors of interest to EFNEP, including food safety, meal planning, use of nutrition labeling, comparing prices, and having children eat breakfast. Aggregate data are available as national summaries, as well as by state and by race—white, black, Hispanic, Native American, and Asian/Pacific Islander. Individual-level data are not currently available for research use, but future plans may include a release of individual-level record data.

The EFNEP data have yet to be used for policy analysis purposes outside of the EFNEP program. Thus, the utility of these data is not fully known. The large samples of low-income individuals and survey questions are potentially valuable for enhancing understanding of policy issues for the low-income population. But since participation in this program is voluntary, participants are not a random sample of the low-income population and are likely to be different from those low-income individuals who did not choose to participate.

5

Recommendations

The need for the U.S. Department of Agriculture (USDA), the U.S Department of Health and Human Services (DHHS), and other agencies to address continuing and emerging policy issues related to food consumption has created new challenges for the available data systems. The ever more pressing need for timely data to help guide policy making places a heavy burden on survey data collections sponsored by the federal government, which are not generally geared toward producing data in a short time frame. In the face of increasing obesity in the United States, there are new calls to understand the economic and social factors behind food consumption and nutrition, creating a need to link data on food consumption to data from various sources on food prices, time use, financial resources, food assistance program participation, availability of food outlets and foods, and other potentially relevant factors. Over the past few decades, changes in the kinds of foods that are consumed and how they are prepared have also posed new challenges for food safety. These changes have increased the need for data on the kinds of foods eaten by specific population groups, such as young children, the elderly, and expectant mothers; the extent to which raw and undercooked foods are consumed; and the extent to which potentially hazardous food additives have been used. Also of concern is the extent of consumer knowledge.

In this chapter, we offer recommendations for improvements in the existing data systems. In formulating recommendations, the panel was asked to focus on improvements that could be made on the margins, within the existing data infrastructure, rather than considering major new data collection efforts. A broader consideration of new data collections would require a much more in-depth study.

INTERAGENCY WORKING GROUP ON FOOD AND NUTRITION DATA

A number of different agencies rely on high-quality data on food consumption, diet, and health. Two departments alone—USDA and DHHS—each have multiple agencies that use these data for different purposes. For some purposes, data on medical and nutritional outcomes and covariates are needed. For others, data about diet and health knowledge and food preparation practices are needed. And for still others, data on prices, expenditures, and financial and other resources are needed. Moreover, for each type of data, there is increasing need for more detail and the ability to link different data sources.

With the merging of the National Health and Nutrition Examination Survey (NHANES) and the Continuing Survey of Food Intakes by Individuals (CSFII) in 2002, NHANES now becomes the only large nationally representative dataset of the federal government that collects detailed information on food consumption. This merger has resulted in new efficiencies in data collection, but it has also placed increased demands on NHANES—to fill both the role that CSFII formerly filled and the primary role of NHANES to assess health and nutritional status.

The process of merging these two data collection programs is still a work in progress. The National Center for Health Statistics (NCHS), the agency in DHHS that conducts the survey, has a good track record of working with other agencies to address data collection needs. But the NHANES will not be able to address all the data needs for all the agencies. Thus, additions and modifications to other related surveys and data collection efforts will need to be considered. In this regard, we are impressed by the initiative shown by the Economic Research Service (ERS) in USDA, not only in funding the development of the Flexible Consumer Behavior Survey Module to include in NHANES and a food and eating module to include in the American Time Use Survey, but also in undertaking research with scanner data and other initiatives to make the best use of and to enhance the available data infrastructure for food and nutrition-related policy planning and analysis.

To build most effectively on the ERS initiative and to take advantage of the current government-wide impetus to deepen understanding of food consumption behavior, its correlates, and its effects on health, safety, and other aspects of society, we recommend an interagency working group on the nation's food and nutrition data infrastructure. In this, we echo and enlarge on a similar recommendation in Dwyer et al. (2003), which reports on a workshop discussion of the integrated NHANES-CSFII.

The proposed interagency group would review the development and collection of new information and make recommendations for design decisions for NHANES and other data sources related to food consumption. It would strive to fill gaps in an effective manner and to reduce unneeded overlaps in data collection. For example, the Food and Drug Administration (FDA) periodically conducts telephone surveys that overlap in content with the kinds of questions on diet and health knowledge and food preparation practices that ERS and NCHS plan to include in the new Flexible Consumer Behavior Survey Module. Perhaps the FCBSM could serve the needs of FDA, or perhaps key question content could be made the same between the FCBSM and relevant FDA surveys so that cross-survey comparisons and validation would be possible.

The proposed interagency group should consider how to develop a complete review of the analytical work that has already been done on assessing the effects of food and nutrition programs in order to identify the information needed to address unanswered questions. It should also consider the kinds of testing and validation that should be built into data collection programs to ensure high-quality information while minimizing respondent burden. It should seek as well to facilitate special arrangements for access to linked datasets that cannot be provided in public-use form.

We believe the group could usefully be led by the Office of Management and Budget Statistical and Science Policy Office, which has a coordinating role in the federal statistical system. Alternatively, it could be co-led by an agency of the USDA working with an agency of DHHS. The group would include representatives from the various agencies in both DHHS and USDA that have policy and data collection responsibilities related to food and nutrition and also from other federal agencies with related policy responsibilities, such as the Environmental Protection Agency.

We understand that interagency working groups are often difficult to make effective because the member agencies have different missions, operations, and cultures. Yet such groups can give visibility to an area, such as food and nutrition policy research, in which coordination and integration of data collection and analysis is needed. Such groups can also provide a venue for systematically considering different perspectives and approaches to collecting quality information in the most efficient and least burdensome manner (see National Research Council, 2005c:12, 44-48).

Recommendation 1: An interagency working group, led by the Office of Management and Budget, or co-led by an agency of the Department of Agriculture and the Department of Health and Human Services, should be established and take responsibility for the systematic development and use of diet and food consumption data to address policy and research questions of the federal government.

RESEARCH AND DEVELOPMENT

To make the proposed interagency working group on diet and food consumption data more effective, we recommend that the group clearly assign lead agency responsibilities for ongoing, sustained research and development programs on data in key areas to inform the group and build a strong base of scientific evidence for its work. Agency research programs should address cost-effective ways to develop high-quality data to remedy data gaps and weaknesses.

For example, ERS could usefully have lead agency responsibility for research to develop high-quality, relevant data to understand the economics of food consumption, factors that affect shopping practices, diet and health knowledge, and related consumer behaviors, and how food-related behaviors affect food consumption and socioeconomic well-being. Such a program should include: assessments of the validity and reliability of alternative datasets; research on linkages of relevant survey data with relevant administrative records, neighborhood characteristics, and retail and household scanner data; and the development of protocols for design and testing of new survey content. Similarly, an agency in DHHS could usefully have lead responsibility for research and development on improved data for monitoring and understanding food fortification or food safety issues.

Recommendation 2: The proposed interagency working group should assign clear responsibilities to lead agencies for sustained programs of research and development on data in key areas to provide a sound base of scientific evidence for the group's work to improve the available information on diet and food consumption.

ENHANCING FOOD AND NUTRITION-RELATED DATA IN NHANES

One of the benefits of previously having two surveys that collected extensive data on food consumption was that each survey could have a different focus. The merged NHANES has not thus far been able to collect all the information that was also on the CSFII, such as information on food expenditures and the information covered by the Diet and Health Knowledge Survey (which was part of the CSFII). Other information, such as data on the relationship between marketing practices, prices, and expenditures for and consumption of food, is also needed.

An important task for the proposed interagency group would be to develop priorities and recommend cost-effective methods for adding food and nutrition-related questions to NHANES. The group could consider alternative methods or designs to obtain additional information—for example, rotating modules or administering one module to half of the survey sample while the other half receives a different module. The group's work would be informed by the ERS research and development program we recommend, which could include the use of small-scale experiments and other methods for testing and validating new survey content in NHANES.

The Flexible Consumer Behavior Survey Module (FCBSM) currently under development by ERS and NCHS will significantly enhance the ability of NHANES to support a wide range of food and nutrition-related research. The module is planned to include questions on food shopping, food expenditures, self-assessment of diet quality, frequency of eating food away from home, attitudes toward and knowledge about diet and food safety, use of food labels, and safety-related food preparation practices. We applaud this effort and urge that it go forward without awaiting the appointment of the interagency working group we recommend. Once appointed, such a group should give the highest priority to reviewing the research and development of the FCBSM and how it can serve the variety of needs for data for food and nutrition-related research and policy analysis.

We note that an important need for analyses of food consumption and shopping and meal preparation behavior is data on actual food prices. This information would likely be difficult to obtain in a supplemental module to NHANES, although it may be possible to ask respondents to scan in some of their purchases; this idea could be tested. It may also be possible to develop linkages with other sources of price information.

Recommendation 3: The proposed interagency working group on diet and food consumption data should consider priorities and methods for obtaining additional food and nutrition-related information in the National Health and Nutrition Examination Survey. The development of the NHANES Flexible Consumer Behavior Survey Module, which will include questions on food expenditures, diet and health knowledge, and other food and nutrition-related topics, should proceed, and research should be conducted on ways to obtain food price information for inclusion in NHANES.

DATA LINKAGES

Our review of existing data sources on food consumption and related information indicates clearly that no single source can satisfy the full range of data needs. Moreover, it does not appear feasible, even if resources were available, to develop a single all-purpose survey on these topics at the level of detail required for many analyses. In addition to the technical and logistical difficulties, the burden on respondents would be too great.

A way to provide needed information at low cost and burden is to look for ways to link data from program administrative records, other surveys, and on-line resources to the NHANES, the Consumer Expenditure Survey, or one of the other datasets (briefly reviewed in Chapter 4) that already contain some relevant food and nutrition-related data. Linkages at the individual level with food assistance program records would provide valuable information on program participation and benefits and associated behavioral effects. Linkages with such sources as area price indexes, census information, and various geographic databases could add metropolitan and neighborhood characteristics that would be helpful for contextual analysis of food consumption behavior. Assessment of the costs, benefits, and methodology for data linkages should be an important component of research and development by ERS in cooperation with other relevant agencies.

Both individual and neighborhood linkages require the ability to geocode the addresses of sample households to small areas. Such linkages also require the implementation of efficient, low-cost means of access to the resulting datasets in ways that protect confidentiality, such as through a research data center, remote on-line access, or a licensing arrangement that permits researchers to use confidential data at their institution.

Recommendation 4: The proposed interagency working group on diet and food consumption data should consider low-cost ways to enhance the analytic uses of NHANES and other surveys by linkages with food assistance program records and with sources of socioeconomic and food shopping characteristics for the areas in which survey respondents live. A priority should be to work out effective ways to provide access to linked datasets through restricted access mechanisms, such as monitored remote on-line access.

USE OF SCANNER DATA

Scanner datasets from retail stores and from household panel scanner surveys include very detailed information on the purchase of specific foods, brands, quantities, and the prices paid. Because they are produced for firms interested in the latest market trends, they are usually available within a matter of weeks or months. This kind of detail and the timeliness with which the data are produced are unmatched in federally sponsored surveys.

Because some policy and decision-making questions require unusually detailed information or must be made with the most timely data, the scanner datasets represent an attractive opportunity for USDA. In particular, retail scanner data have the potential for understanding short- and long-term trends in aggregate market purchasing of foods. Both these data and the household panel scanner datasets can be used to understand how consumer behavior changes (on both a very short-term and long-term basis) with respect to specific product attributes and price, when new products are introduced, or when labeling regulations change. Proprietary data on household food consumption also have the potential to identify trends in eating patterns and dieting practices.

ERS has been exploring the use of these proprietary datasets for several years. It has contracted with ACNielsen for data from the HOMESCAN panel and with the NPD Group for the National Eating Trends data. However, the quality of these data is largely untested. Data quality concerns include the representativeness of the samples of the U.S. population overall and of certain groups, such as low-income, single-adult, and minority households. For scanner data, accuracy of coding and the accuracy of the product scanning process are also of concern. Moreover, the retail scanner data do not cover all retailers, food that does not have a universal product code (UPC), or food purchased in restaurants.

It would be beneficial for ERS to further explore the use of these data. Because scanner data are proprietary, it is not possible to make them readily available at low cost to all agencies and researchers that could benefit from using them. ERS could, however, explore with market research firms ways to obtain older scanner datasets that are of less value for the firms' private-sector clients at a favorable price for redistribution to other federal users. For example, studies that compare food purchases from these datasets with data on food purchases from the Consumer Expenditure Survey (CE) and with food consumption data from the NHANES dietary intake survey could be conducted. Studies on the quality of the data obtained through the scanners could also be conducted. Ways to link scanner data at the neighborhood level with the NHANES or the Consumer Expenditure Survey could be explored as well.

As a first step in this area, ERS should consider holding a conference of policy analysts and researchers who have used scanner and related proprietary data to see what has changed since its 2003 conference (see Chapter 3). The conference should address funding and research priorities in three areas: research on data quality (what is known and what is needed); accessibility of the data and how cost and access barriers can be reduced; and what research and policy questions can benefit most from scanner data.

Recommendation 5: The Economic Research Service of the U. S. Department of Agriculture should continue to explore the use of data on food purchases, prices, and consumption from proprietary retail scanner systems, household scanner surveys, and household consumption surveys. This work should include a program to examine the quality of the data, consideration of ways to reduce the costs of access, and the determination of priority applications for the information.

USE OF OTHER DATASETS

Our review has focused principally on the major surveys for food consumption analysis, including the National Health and Nutrition Examination Survey, the Continuing Survey of Food Intakes by Individuals (including the Diet and Health Knowledge Survey supplement to the CSFII), the Consumer Expenditure Survey, and scanner datasets and household food consumption data collected by market research firms. There are many other federal datasets that, while they primarily serve other purposes, include some relevant information and could be useful for food and nutrition-related policy analysis and research with modest enhancements. We urge the proposed interagency working group, informed by research and development by the Economic Research Service and other relevant agencies, to consider low-cost ways to exploit surveys such as the Current Population Survey, the American Time Use Survey, panel surveys of specific age groups or the low-income population, and surveys that are designed for the addition of modules to track emerging trends.

Recommendation 6: The proposed interagency working group on diet and food consumption data should consider ways to enhance the usefulness of other federal datasets for food and nutrition-related policy analysis and research. Such datasets include the Current Population Survey, the American Time Use Survey, panel surveys that follow families, children, and the elderly over time, and surveys that are designed to include modules to track emerging trends.

CONCLUSION

This report has reviewed the kinds of information and data needed to more fully understand decisions that the population makes on food consumption and to guide policy makers. We believe the implementation of our recommendations and consideration of the suggestions we make throughout the report will improve the underlying knowledge base for food and nutrition-related policy planning in the United States.

References

Abraham, K. (2004). *Learning About How Americans Spend Their Time*. Prepared for Food and Eating Consequences of Time-Use Decisions: A Research and Policy Conference. Economic Research Service, U.S. Department of Agriculture, Washington, DC [July 13].

Aten, B.H. (2005). *Report on Inter-Area Price Levels*. Joint Research with the Office of Price and Living Conditions, Bureau of Labor Statistics. Regional Economics Directorate, Bureau of Economic Analysis, Washington, DC [May 18].

Blisard, N., H. Stewart, and D. Jolliffe (2004). *Low-Income Households' Expenditures on Fruits and Vegetables*. Agricultural Economic Report Number 833. Washington, DC: U.S. Department of Agriculture.

Blisard, N., J.N. Variyam, and J. Cromartie (2003). *Food Expenditures in U.S. Households: Looking Ahead to 2020*. Agricultural Economic Report No. 821. Washington, DC: Economic Research Service, U.S. Department of Agriculture [February].

Bonnet, C., and M. Simioni (2001). Assessing consumer response to protected designation of origin labeling: A mixed multinomial logit approach. *European Review of Agricultural Economics* 28(4):433-449.

Bureau of Labor Statistics (2004a). *American Time Use Survey User's Guide 2004*. Washington, DC: Bureau of Labor Statistics and U.S. Census Bureau.

——— (2004b). *Consumer Expenditures in 2002*. BLS Report 974, U.S. Department of Labor [February]. Available: http://www.bls.gov/cex/csxann02.pdf [June 2005].

——— (2004c). *Time-Use Survey—First Results Announced by BLS*. News release USDL 04-1797. Washington, DC: U.S. Department of Labor [September 14].

Chevalier, J.A., A.K. Kashyap, and P.E. Rossi (2004). Why don't prices rise during periods of peak demand? Evidence from scanner data. *American Economic Review* 93(1):15-37.

Chou, S., M. Grossman, and H. Saffer (2004). An economic analysis of adult obesity: Results from the Behavioral Risk Factor Surveillance System. *Journal of Health Economics* 23(3):565-587.

Cody, S., and C. Tuttle (2002). *The Impact of Underreporting in CPS and SIPP on Microsimulation Models and Participation Rates.* Memo No. 483. Mathematica Policy Research, Inc., Washington, DC [July 24].

Cole, N. (1997). *Evaluation of the Expanded EBT Demonstration in Maryland: Food Store Access and Its Impact on the Shopping Behavior of Food Stamp Households.* Prepared for the Food and Nutrition Service, U.S. Department of Agriculture. Abt Associates, Inc.

——— (2003). *Feasibility and Accuracy of Record Linkage to Estimate Multiple Program Participation: Volume I, Record Linkage Issues and Results of the Survey of Food Assistance Programs.* Prepared for the Economic Research Service, U.S. Department of Agriculture. Abt Associates, Inc. [June].

Cole, N., and E. Lee (2004). *Feasibility and Accuracy of Record Linkage to Estimate Multiple Program Participation: Volume III: Results of Record Linkage.* Prepared for the Economic Research Service, U.S. Department of Agriculture. Abt Associates, Inc. [November].

Cotterill, R.W., and A.W. Franklin (1999) An estimation of consumer benefits from the public campaign to lower cereal prices. *Agribusiness: An International Journal* 15(2):273-287.

Cutler, D.M., E.L. Glaeser, and J.M. Shapiro (2003). Why have Americans become more obese? *Journal of Economic Perspectives* 17(3):93-118.

Day, K., B. Kuhn, and A. Vandemann (1995). Measuring the food safety risk of pesticides. In *Valuing Food Safety and Nutrition,* J.A. Caswell, ed. Boulder, CO: Westview Press.

Dong, L., G. Block, and S. Mandel (2004). Activities contributing to total energy expenditure in the United States: Results from the NHAPS study. *International Journal of Physical Activity* 1(1):4.

Dwyer, J., M.F. Picciano, D.J. Raiten, and Members of the Steering Committee (2003). Collection of food and dietary supplement intake data: What We Eat in America-NHANES. *Journal of Nutrition* 133:590S-600S (February).

Federation of American Societies for Experimental Biology (1995). *Third Report on Nutrition Monitoring in the United States.* Washington, DC: U.S. Government Printing Office.

Fox, M.K., W.L. Hamilton, and B.-H. Lin (2004a). *Effects of Food Assistance and Nutrition Programs on Nutrition and Health: Volume 3, Literature Review.* Food Assistance and Nutrition Research Report Number 19-3 Washington, DC: U.S. Department of Agriculture [December].

——— (2004b). *Effects of Food Assistance and Nutrition Programs on Nutrition and Health: Volume 4, Executive Summary of the Literature Review.* Food Assistance and Nutrition Research Report No. 19-4. Washington, DC: U.S. Department of Agriculture [December].

French, S.A., L. Harnak, and R.W. Jeffey (2000). Fast food restaurant use among women in the pound of prevention study. *International Journal of Obesity* 24(10):1353-1359.

French, S.A., M. Storey, D. Neumark-Sztainer, J.A. Filkerson, and P. Hamman (2001). Fast food restaurant use among adolescents. *International Journal of Obesity* 25:1823-1833.

Frost, A. (2004) *What are the Policy Questions?* Prepared for Food and Eating Consequences of Time-Use Decisions: A Research and Policy Conference. Economic Research Service, U.S. Department of Agriculture, Washington, DC [July 13].

Garner, T., G. Janini, W. Passero, L. Paszkiewicz, and M. Vendemia (2003). *The Consumer Expenditure Survey in Comparison: Focus on Personal Consumption Expenditures.* Prepared for the Federal Economic Statistics Advisory Committee Meeting. Bureau of Labor Statistics and U.S. Census Bureau, Washington, DC [March 17].

Haines, P.M., M.Y. Hama, D.K. Guilkey, and B.M. Popkin (2003). Weekend eating in the United States is linked with greater energy, fat, and alcohol intake. *Obesity Research* 11:945-949.

Hamermesh, D. (2004). *Ten Research Questions You Might Answer with ATUS Food-Related Data*. Prepared for Food and Eating Consequences of Time-Use Decisions: A Research and Policy Conference. Economic Research Service, U.S. Department of Agriculture Washington, DC.

Hamilton, W.L., and P.H. Rossi (2002). *Effects of Food Assistance and Nutrition Programs on Nutrition and Health: Volume 1, Research Design*. Food Assistance and Nutrition Research Report Number 19-1. Washington, DC: U.S. Department of Agriculture [February].

Hamrick, K. (2004). *The Food and Eating Module of the American Time Use Survey*. Prepared for Food and Eating Consequences of Time-Use Decisions: A Research and Policy Conference. Economic Research Service, U.S. Department of Agriculture, Washington, DC.

Herz, D. (2004). *The American Time Use Survey*. Prepared for Food and Eating Consequences of Time-Use Decisions: A Research and Policy Conference. Economic Research Service, U.S. Department of Agriculture, Washington, DC.

Hurd, M., and Rohwedder, S. (2003). *The Retirement- Consumption Puzzle: Anticipated and Actual Declines in Spending at Retirement*. NBER Working Paper No. 9586. National Bureau of Economic Research, Cambridge, MA.

Institute of Medicine (2004). *Proposed Criteria for Selecting the WIC Food Packages: A Preliminary Report of the Committee to Review the WIC Food Packages*. Washington, DC: The National Academies Press.

——— (2005). *Preventing Childhood Obesity—Health in the Balance*. Committee on Prevention of Obesity in Children and Youth, Food and Nutrition Board, J.P. Kaplan, C.T. Liverman, and V.I. Kraak, eds. Washington, DC: The National Academies Press.

Jacobs, E., and S. Shipp (1990). A history of the U.S. Consumer Expenditure Survey: 1935 to 1988. In *Proceedings of the Social Statistics Section*. Alexandria, VA: American Statistical Association.

Jensen, H. (2003). *Demand for Enhanced Foods and the Value of Nutritional Enhancements of Food*. Prepared for a Workshop on the Use of Scanner Data in Policy Analysis. Economic Research Service, U.S. Department of Agriculture, Washington, DC [June 9].

Jones, E. (1997). An analysis of consumer food shopping behavior using supermarket scanning data: Differences by income and location. (Income Inequality: Implications for Food Consumption Behavior). *American Journal of Agricultural Economics* 79:1437-1443.

Kirlin, J., and N. Cole (1999). *Feasibility Study of Capturing Food Data at Checkout*. Alexandria, VA: Food and Nutrition Service, U.S. Department of Agriculture.

Kokoski, M., B. Moulton, and K. Zieschang (1999). Interarea price comparisons for heterogeneous goods and several levels of commodity aggregation. Pp. 123-166 in *International and Interarea Comparisons of Income, Output, and Prices*, A. Heston and R. Lipsey, eds. Chicago: University of Chicago Press.

Kuchler, F., K. Ralston, and L. Unnevehr (1997). Reducing pesticide residue risks in consumers: Can agricultural research help? *Food Policy* 22(2):119-132.

Lakdawalla, D.N., and T.J. Philipson (2002). *Technological Change and the Growth of Obesity.* Cambridge, MA: National Bureau of Economic Research.

Lino, M., J.M. Dinkins, and L. Bente (1999). Household expenditures on vitamins and minerals by income level. *Family Economics and Nutrition Review* 12(2):39-43.

Logan C., M.K. Fox, and B.-H. Lin (2002). *Effects of Food Assistance and Nutrition Programs on Nutrition and Health: Volume 2, Data Sources.* Food Assistance and Nutrition Research Report Number 19-2. Washington, DC: U.S. Department of Agriculture [September].

MacDonald, J.M. (2000) Demand, information, and competition: Why do food prices fall at seasonal demand peaks? *Journal of Industrial Economics* 48(1):27-45.

Mathios, A.D. (1998). The importance of nutrition labeling and health claim regulations on product choice: An analysis of the cooking oil market. *Agricultural and Resource Economics Review* 27(2):159-168.

—————— (2000). The impact of mandatory disclosure laws on product choices: An analysis of the salad dressing market. *Journal of Law and Economics* 43(2):651-77.

Maynard, L.J., and V.N. Veeramani (2003). Price sensitivities for US frozen dairy products. *Journal of Agricultural and Applied Economics* 35(3):599-609.

Meyer, Bruce D., and James X. Sullivan (2004). *Consumption and the Poor: What We Know and What We Can Learn.* Prepared for the ASPE-Initiated Workshop on Consumption Among Low-Income Families. University of Chicago and University of Notre Dame [October 25].

Mladenic, D., W.F. Eddy, and S. Ziolko (2001). *Exploratory Analysis of Retail Sales of Billions of Items.* Proceedings of the 33rd Symposium on the Interface. Available: http://www.galaxy.gmu.edu/interface/I01/I2001Proceedings/WEddy/WEddy.pdf [June 2005].

National Research Council (1995). *Measuring Poverty: A New Approach.* Panel on Poverty and Family Assistance, Committee on National Statistics, C.F. Citro and R.L. Michael, eds. Washington, DC: National Academy Press.

—————— (1999). *Evaluating Food Assistance Programs in an Era of Welfare Reform: Summary of a Workshop.* Committee on National Statistics and Institute of Medicine, E. Evanson, C.F. Manski, and T.M. Scanlan, eds. Washington, DC: National Academy Press.

—————— (2003). *Frontiers in Agricultural Research: Food, Health, Environment, and Communities.* Committee on Opportunities in Agriculture. Washington, DC: The National Academies Press.

—————— (2004). *Exploring a Vision: Integrating Knowledge for Food and Health.* Board on Agriculture and Natural Resources, T.I. Rouse and D. P. Davis, eds. Division of Earth and Life Studies. Washington DC: The National Academies Press.

—————— (2005a). *Expanding Access to Research Data: Reconciling Risks and Opportunities.* Panel on Data Access for Research Purposes, Committee on National Statistics. Washington, DC: The National Academies Press.

—————— (2005b). *Measuring Food Insecurity and Hunger—Phase 1 Report.* Panel on USDA's Measurement of Food Insecurity and Hunger, Committee on National Statistics. Washington, DC: The National Academies Press.

——— (2005c). *Principles and Practices for a Federal Statistical Agency*, 3rd edition. Committee on National Statistics, M.E. Martin, M.L. Straf, and C. F. Citro, eds. Washington, DC: The National Academies Press.

Nord, M., and G. Bickel (2002). *Measuring Children's Food Security in U.S. Households, 1995-99*. Food Assistance and Nutrition Research Report No. 25. Washington, DC: Economic Research Service, U.S. Department of Agriculture [April].

Nord, M., M. Andrews, and S. Carlson (2004). *Household Food Security in the United States, 2003*. Food Assistance and Nutrition Research Report Number 42. Washington, DC: Economic Research Service, U.S. Department of Agriculture.

Ohls, J., M. Pnoza, L. Moreno, A. Zambrowski, and R. Cohen (1999). *Food Stamp Participants' Access to Food Retailers*. Final Report to the Food and Nutrition Service, U.S. Department of Agriculture. Princeton, NJ: Mathematica Policy Research, Inc.

Paulin, G. D. (1995). *The Changing Food at Home Budget: 1980 and 1992 Compared*. BLS Paper. Bureau of Labor Statistics, Washington, DC. Available: http://www.bls.gov/cex/csxewp.htm [June 2005].

Putsis, W.P., and R.W. Cotterill (2001) Do models of vertical strategic interaction for national and state brands meet the market test? *Journal of Retailing* 77(1):83-109.

Reed, J., E. Frazao, and R. Itskowitz (2003). *How Much Do Americans Pay for Fruits and Vegetables?* Agriculture Information Bulletin No. 790. Washington, DC: Economic Research Service, U.S. Department of Agriculture.

Reinsdorf, M. (1999). Using scanner data to construct basic component indexes. *Journal of Business and Economic Statistics* 17(2):152-160.

Taeuber, C., D. M. Resnick, S. P. Love, J. Staveley, P. Wilde, and R. Larson (2004). *Differences in Estimates of Food Stamp Program Participation Between Surveys and Administrative Records*. Washington, DC: U.S. Census Bureau.

Thompson, G.D., and P.N. Wilson (1997). The organizational structure of the North American fresh tomato market: Implications for seasonal trade disputes. *Agribusiness: An International Journal* 13:533-547.

Unnevehr, L.J., and F.C. Gouzou (1998). Retail premiums for honey characteristics. *Agribusiness: An International Journal* 14(1): 49-54.

U.S. Department of Agriculture (2000). *Results from USDAs 1994-96 Diet and Health Knowledge Survey: Table Set 19*. ARS Food Surveys Research Group, Agricultural Research Service, Washington, DC. Available: http://www.barc.usda.gov/bhnrc/foodsurvey/home.htm [June 2005].

——— (2001). *Draft Risk Assessment of the Public Health Impact of Escherichia coli O157:H7 in Ground Beef*. Food Safety and Inspection Service, Washington, DC. Available: http://www.fsis.usda.gov/OPPDE/rdad/FRPubs/00-023N/00-023NReport.pdf [June 2005].

Vickner, S.S., and S.P. Davies (2002). Estimating strategic price response using cointegration analysis: The case of the domestic black and herbal tea industries. *Agribusiness: An International Journal* 18(2):131-144.

Ward, M.B., J.P Shimshack, J.M. Perloff, and J.M. Harris (2002). Effects of the private-label invasion in food industries. *American Journal of Agricultural Economics* 84:961-973.

West, J., K. Denton, and L.M. Reaney (2000). *The Kindergarten Year: Findings from the Early Childhood Longitudinal Study, Kindergarten Class of 1998-1999*. NCES Publication 2001023. National Center for Education Statistics, U.S. Department of Education, Washington, DC. Available: http://nces.ed.gov/pubs2001/2001023.pdf [June 2005].

Appendix A

Enhancing the Data Infrastructure in Support of Food and Nutrition Programs, Research, and Decision Making: Summary of a Workshop

As part of its data-gathering activities, the Panel on Enhancing the Data Infrastructure in Support of Food and Nutrition Programs, Research, and Decision Making hosted a workshop in Washington, D.C., on May 27-28, 2004. The workshop served as a forum for input from agencies with policy responsibilities, agencies that provide relevant data, private firms that produce relevant data, and independent researchers. (See Appendixes B and C for the workshop agenda and participants.)

Representatives from the U.S. Department of Agriculture (USDA) and from the Food and Drug Administration (FDA), the National Institutes of Health (NIH), and the Environmental Protection Agency (EPA) discussed current and emerging data needs related to food consumption for policy and decision making. Representatives from key federal statistical agencies that produce food consumption and expenditures datasets—the National Center for Health Statistics (NCHS) of the Centers for Disease Control and Prevention (CDC) and the Bureau of Labor Statistics (BLS)—discussed the strengths and limitations of currently available data. Representatives from private firms that produce data on food consumption and expenditures, the NPD Group and ACNielsen, also discussed current data and data gaps. Outside researchers responded to the presentations and suggested possible improvements to the data infrastructure. The workshop was geared toward national data only, so large state-level dietary databases were not discussed.

This appendix summarizes the workshop. The next four sections cover the sessions at which presenters from federal agencies, private firms, and academia discussed existing data and data needs in four areas. The last two sections summarize the issues and questions raised by members of the panel and others on two forward-looking topics: use of proprietary data for policy purposes and possible data improvements.

This summary does not offer any conclusions or recommendations; those are in the main body of the panel's report (Chapter 5). The panel's mission in this workshop and in its deliberations was to consider modest data improvements that could be made to the current data infrastructure with little expense, such as adding new questions to existing surveys and linking existing datasets. The panel was not asked to consider a major overhaul of data systems.

Many topics were covered during the one-and-a-half day workshop. The workshop began with an overview of food consumption, expenditures, and sales datasets, focusing on the National Health and Nutrition Examination Survey, the Consumer Expenditure Survey, and proprietary datasets from food market research firms. The next three sessions of the workshop were devoted to the food consumption and expenditure data needs for different agencies in USDA and in other federal agencies: one session focused on food marketing and promotion and food market analysis at which there were presentations of the food consumption data needs of USDA's Agricultural Marketing Service and a description of USDA's World Agricultural Board's economic forecasts; one focused on food consumption data and the evaluation of food assistance programs for monitoring and evaluation; and one covered food safety and food consumption data and featured presentations describing current uses of food consumption data by USDA's Food Safety and Inspection Service, FDA's Office of Food Additive Safety, and the Environmental Protection Agency's Office of Pesticide Programs. The fifth session, on food consumption data and health, included presentations from USDA's Center for Nutrition Policy and Promotion and from the National Cancer Institute. Each of these sessions included a discussion of data needs for that topic by individual researchers working outside of the federal government. The sixth session was devoted to the use of proprietary sources of data to address policy questions and included presentations from representatives of ACNielsen and the NPD Group, along with a presentation about the applications of these data. The final workshop session consisted of a panel discussion on possible data improvements or data linkages, with participation by four panel members. The topics of supplement intakes and food composition databases were discussed only briefly at the workshop.

OVERVIEW OF FOOD CONSUMPTION, EXPENDITURES, AND SALES DATASETS

During this session, presenters reviewed two public datasets, the National Health and Nutrition Examination Survey (NHANES) and the Consumer Expenditure Survey (CE), and data produced by market research firms. NHANES is a major source of food consumption data, but the NPD Group's National Eating Trends (NET) also gathers food consumption data. Food expenditure data offer another view into food behavior. The CE and proprietary scanner data offer data on household food expenditures. This session of the workshop consisted of overviews of these datasets by Clifford Johnson of NCHS on NHANES, Steven Henderson of BLS on CE, and Abebayehu Tegene from ERS on proprietary sources of data.

Nutritional Component of NHANES

The National Health and Nutrition Examination Survey (NHANES) is a cross-sectional survey conducted by NCHS. NHANES' objective is to assess the health and nutritional status of adults and children in the United States, and the data are used in making public health policy. Clifford Johnson, director of the NHANES program at NCHS, gave an overview of NHANES specifically focusing on its nutrition component. NHANES data collection began in 1971 and became an annual survey in 1999. Information is collected from about 5,000 participants of all ages annually. Participant information is pooled for every 2 years of data collection. Topics covered by NHANES range from mental health to obesity to bone density to environmental exposures (see http://www.cdc.gov/nchs/nhanes.htm). In addition to obtaining information on respondents' demographic background, basic medical, anthropological, and health status information is obtained. This information is collected when respondents visit Mobile Examination Centers (MECs), which travel around the country to administer the surveys.

NHANES nutrition assessments include dietary nutrient intake; anthropometric measurements; nutritional biochemistries and hematologic tests; physical examination; and interview. Anthropometric measurements include height, weight, and body mass index (BMI). Nutritional biochemistries and hematologic tests include measures of iron and folate status as well as other vitamins, minerals, electrolytes, cholesterol, and triglycerides. NHANES also includes body composition measurements. The physical fitness assessment considers both daily living activities, such as walking to work or house and yard work, and leisure physical activities, such as muscle-strengthening activities and sedentary activities, including watching television.

The NHANES 1999-2001 dietary assessment included a 24-hour recall of food intake for all participants. Respondents were asked to record everything they ate and drank for 24 hours. A subset of persons was asked for a second day of recall. The day two sample was selected to have representative coverage of the full sample's age/sex subgroups, but it may not have been fully representative of the total population because it only included about 8 percent of the original sample. During an interview, focused dietary questions were asked, along with detailed questions about supplements and medications and food security. The dietary behavior assessment asked questions about alcohol consumption, salt use at the table, and the frequency of consumption of vegetables, fish and shellfish, and skin on chicken and visible fat on meat. Other questions covered self-reported weight during a person's life, self-perception of weight, and weight control practices.

NHANES was changed in 2002. Until 2001, the Agricultural Research Service of the USDA had conducted the Continuing Survey of Food Intakes by Individuals (CSFII). The last year of data collection for the CSFII was in 1996 for adults and 1998 for children. In order to reduce redundancy in national dietary data collection, CSFII was discontinued, and the USDA staff responsible for the CSFII began working on the dietary component of NHANES, beginning with the 2002 data collection. The dietary component of NHANES is now called "What We Eat in America," and it includes 2 days of dietary intake data. The first day of the recall is done by an in-person interview in a mobile examination center, and the second day is done by telephone interview in the participant's home. CSFII also contained the Diet and Health Knowledge Survey (DHKS), which studied respondents' dietary knowledge and so provided information on why people choose certain foods and beverages. James Blaylock of ERS noted that the President's 2005 budget includes a data initiative that will reinstate a diet and health knowledge survey. According to a workshop participant, these questions will add 5-7 minutes to NHANES. The original DHKS lasted 20-30 minutes by telephone, so the questions will be modified. Ronette Briefel also noted that the FDA is currently testing the Health and Diet Survey, which will be similar to the DHKS, but will not include any sort of dietary recall as does NHANES.

Beginning with NHANES 2003-2004, physical activity monitors have been used on participants aged 6 and older. Each participant is asked to wear a monitor for 7 days and then return it by mail to the researchers. The monitor measures the intensity and duration of locomotion activities and number of steps taken. A food frequency questionnaire, known as the Food Propensity Questionnaire (FPQ), will be used for participants aged 2 and over. The FPQ gathers information about food consumption probabilities on a given day, includes 134 questions on individual food items and food groupings, and helps in estimating usual intake for those foods. The FPQ was adapted from the National Cancer Institute's (NCI) Diet Health Questionnaire. NCI collaborated with NHANES to develop and field a pilot test of the FPQ (see http://riskfactor.cancer.gov/studies/nhanes).

Consumer Expenditure Survey

Steven Henderson from BLS gave a presentation on the Consumer Expenditure Survey (CE). The CE comprises two surveys—a diary survey, in which households keep an expenditure diary, and a quarterly household interview survey, which collects information on major purchases on a quarterly basis and background information on consumers. Information is collected continuously from 105 geographic areas of the United States. The diary survey is a record of daily expenses for a consumer unit that is kept by a respondent from each consumer unit for two consecutive 7-day periods. Each quarter, 7,500 consumer units participate in the interview, and 7,500 consumer units annually complete two diaries. U.S. Census Bureau interviewers explain to respondents how to fill out the diary and review the diaries for completeness when they collect them at the end of each week.

All daily expenses, except business expenses and expenses incurred while out of the home overnight, are included in the diary. It also collects demographic, work experience, and income data on household members aged 15 and over. The household interview survey also includes questions to double check food purchases reported in the diary portion of the survey, including questions about food purchased away from home. Data from the two surveys are integrated to provide information about both detailed day-to-day purchases and long-term, major purchases.

The CE annual data are usually available one year after they are collected in ten standard tables sorted by key demographic variables. BLS also produces unpublished tables, which are available on request, that contain more expenditure detail. For example, instead of a general "beef" category, the unpublished table includes categories for ground beef, chuck roast, round roast, other roast, round steak, sirloin steak, other steak, and other beef. These tables are unpublished due to the higher variance associated with a more limited sample of persons making these expenditures. The CE provides the market basket of weights for the Consumer Price Index (CPI) and an annual snapshot of all spending by key demographic variables for researchers, analysts, and agencies.

BLS provides microdata in a CD format. The microdata include detailed food purchases reported in dollars (not quantity). Summary food statistics combine subcategories, such as fresh vegetables and fresh fruits. Demographics of household members include income, education, age, gender, race, home ownership, job status, plus Food Stamp Program participation, free meals, and poverty status. The CE does not collect data on who in the family made the purchase, who consumed the food, and the quantity purchased.

Proprietary Data

James Blaylock from ERS noted that public national datasets offer rich and important data, but they take a long time to produce and analyze.[1] Because of their relationships with retailers and their customer's needs and willingness to pay for the absolutely latest trend information, datasets produced by marketing firms are produced on a more timely basis. These data typically include information about what food people buy, what they eat, and demographic information about households. Such data are a potential resource to the USDA to fill gaps in the current data infrastructure. Abebayehu Tegene from ERS gave an informative presentation on proprietary data. He discussed both scanner data, which provide purchase information, and survey data, which provide consumption information.

Scanner data come from two types of data collections: *point-of-sale* collections, which use the universal product codes (UPC) of products sold in retail checkout counters to identify products and quantities sold and their prices; and *household scanner panels*, which are usually random samples of households for which members are asked to scan in the items they have

---

[1]For example, NHANES data gathered in 1999-2000 began to be released in June 2002. One workshop participant complained that some dietary data for NHANES 1999-2000, such as the recipe files, had not been released as of May 2004.

purchased. Household panel scanner data are generally gathered from large samples. For example, ACNielsen has a household panel of 61,500. The samples are selected randomly on the basis of demographic and geographic targets. (Statistical weighting perhaps can expand the sample to the U.S. level.) The household panels provide information on who buys what and where; the data are gathered weekly; see Box 3-1 in Chapter 3 for the information Tegene noted was gathered by ACNielsen and IRI. There is a 12-day to 3-week lag between data collection and release, depending on the vendor.

The NPD Group and ACNielsen are two companies that provide survey data. The NPD Group has two ongoing surveys: the Consumer Report on Eating Share Trends (CREST) and the National Eating Trends (NET). CREST tracks consumer purchases of prepared meals and snacks at commercial restaurants with an online panel of 52,500. It is geared towards food service information and includes what is eaten, where and with whom, and how much is spent. It contains sales information by food type and outlet. Every day 3,500 questionnaires are sent—3,000 to adults and 500 to teens. The questionnaires ask for information for the day before. CREST represents each of the nine Census Bureau regions. Data are gathered monthly and are available 1 month after collection.

NET collects information on food and beverage consumption for all family members of 2,000 households during a 2-week period, using 14 daily paper diaries. Food and beverage consumption is recorded for each household member for 14 consecutive days. Questionnaires are mailed to 3,500 households that are selected from NPD's panel of more than 50,000 households, with data collected continuously throughout the year. NET data are weighted by five key demographic variables: income, family size, age, employment status, and race. Data are available 3 months after collection. The diary does not collect information on food prices, but it does collect in-depth information about what and how much is consumed, who consumes it, and when and where food is consumed. The diary also collects information on diet status and type, height and weight, vitamin and mineral consumption, exercise habits, and nutritional attitudes. NET uses USDA's information on nutrient composition and serving sizes to convert the collected data to food pyramid food groups.

Tegene noted that proprietary data have some shortcomings for some purposes. There is some concern about the sampling frames used to collect the data and how well they represent populations. Some survey data are gathered through the Internet, to which not all households, especially low-income households, have access. Tegene also said that it is important to note that scanner data are not a direct measure of food consumption because they only include information about store purchases and only address purchase behavior, not consumption.

Helen Jensen of Iowa State University also offered some criticisms of the Nielsen HomeScan Panel during her workshop presentation. Members of the sample have higher incomes and smaller household sizes than the general population, and they are more likely to be married, more likely to be white, and less likely to be Hispanic. Jensen noted that when working with scanner data, it is important to think about whether the sample is representative and if low-income and minority populations are included in sufficient numbers. It is also important to think about whether store purchases are representative: purchases from convenience stores and other small stores may not be included. CREST does not include data on vending machine, bar and tavern, and social catering consumption behavior. It is also important to know if food assistance program purchases can be identified.

The household survey data also do not include information on the price a consumer paid for the goods. Proprietary data, especially very current data, are also more expensive to obtain than public data. Some researchers expressed concern that proprietary data will never be publicly available in tables or other raw form. While NHANES data are available for free download from the Internet in various forms, such as tables or microdata that have been protected for confidentiality, workshop participants questioned whether the firms will ever allow USDA to share those data publicly. There are also methodological issues with self-administered data: they are less accurate than interviewer-administered surveys because people may not completely understand what information to provide.

Despite their shortcomings, proprietary data could be very useful. Tegene and Blaylock noted that they are timely and detailed, providing current information on who purchases and consumes what and where. Blaylock used the case of mad cow disease in the United States as an example. If USDA had had access to private datasets, it would have been able to add questions to a private survey within days of the first news stories. The government would then have known almost immediately how consumers' purchases and consumption of beef changed in response to the outbreak.

Proprietary data combine detailed product and household information, so researchers can study both retailer and consumer behavior. Proprietary data lend themselves to different analytic approaches, which allow them to address emerging issues. For example, panel demographic and purchase behavior data can be used to address specific research and policy issues. Tegene noted that these data could be used to answer policy questions about obesity, such as: Do taxes on less healthy foods or subsidies on healthy foods change people's purchases and consumption? What is the role of advertising in children's obesity? Vendors could also be asked to provide focused surveys, like surveys on teenagers' consumption behavior. They could also provide customized reports studying specific products or subgroups of items in greater detail.

During the workshop sessions, representatives of different USDA agencies and other federal agencies were asked to discuss key and emerging policy questions related to food consumption and data needed to answer those policy questions. The next four sections summarize these data needs.

FOOD MARKETING AND PROMOTION AND FOOD MARKET ANALYSIS

Food Consumption Data in Regulatory Analysis and Generic Promotion

Don Hinman of USDA's Agricultural Marketing Service (AMS) began his presentation with a description of the agency's focus on its food-related programs. AMS has both regulatory functions and marketing services: it establishes grade standards, provides grading and inspection services for a fee, provides market news and dairy marketing orders, and sets minimum milk prices; it also purchases commodities for school lunch and other feeding programs and provides oversight of generic promotion programs.

AMS uses consumption data to learn what consumers are looking for and to determine the effect of grade changes on purchases. AMS has an interest in having good consumption data available to researchers and for its own use in analysis of regulatory effects and marketing services. AMS also uses consumption data to analyze the effects of regulations and marketing orders. Panel data that measure purchase decisions in relation to quality attributes, such as fruit maturity, could facilitate better quantitative economic analysis. Consumption data can inform AMS when a commodity industry is in distress due to increased production or large inventories. They can also inform AMS if commodity demand is stagnant or declining. Consumption data that include preferences for food attributes are useful for analyzing economic effects. For example, does country-of-origin labeling affect food purchases? To what extent are organic products a distinct market? What is the influence of USDA's organic seal on consumer purchasing behavior?

AMS is most interested in per capita consumption data. This key variable is used in economic models designed to measure promotional effects (the way in which consumption changes after an advertising campaign, for example), so maintaining or improving the quality of per capita consumption data thus is critical to the agency. It is very helpful to distinguish where food is consumed—at home or away from home.

AMS used proprietary panel data from the NPD Group for an informative project on beef consumption. The project looked at servings of beef consumed per household member in a 2-week period. Information on the effects of prices, demographics, health and diet concerns, and the effect of promotions on the likelihood and amount of beef consumption were gathered. The unique panel data allowed AMS to analyze consumption before and after the promotion. These data are then used to estimate a rate of return to producers on promotional dollars expended. Beef promotion led to an increase in beef servings by 0.20 serving per household member. The estimated additional amount consumed due to the promotion was about 2 ounces per household member. The project found that more concern with cholesterol was associated with lower beef consumption and more fast food purchasing was associated with higher beef consumption.

One concern expressed with the use of such proprietary data was losing access to it. AMS has already ceased using the NPD Group's eatings dataset because of high costs. NPD Group's servings dataset may also become cost prohibitive. Researchers working on commodity promotion effects value future public investment in panel data and other sources of commodity-specific data on food consumption.

Demand Forecasts and the
World Agricultural Supply and Demand Estimates Report

The World Agricultural Outlook Board (WAOB) is charged with coordinating the interagency process for preparing USDA's economic forecasts of commodity supply and use. WAOB analysts must be knowledgeable about the latest trends in food consumption and the factors that affect the use of commodities. The WAOB's George Bange began his presentation with a discussion of fundamentals. Commodity analysis uses a supply-and-demand framework in which the total supply amount (carry-in stocks, production, and imports) and total disappearance amount (exports, domestic use, consumption, spoilage, and carry-out stocks) must equal. Accurate forecasts of domestic disappearance amounts, which are the dominant use category for most commodities, are essential.

Analysis of observed prices for agricultural commodities indicates that short-term changes in supply are typically inversely correlated with prices. In the short term, WAOB tracks changes along a demand curve. Shifts in demand occur over a longer time period, and identifying these shifts is difficult without sound knowledge of underlying factors, such as income, social and demographic characteristics, changes in animal feeding practices and feed technology, government policies, and dietary preferences.

USDA and WAOB project consumption or food use on both a short-term and a long-term basis. USDA publishes the *World Agricultural Supply and Demand Estimates* (*WASDE*) report on a monthly basis. This report has many users: farmers use it to project prices and market conditions; the Chicago Board of Trade uses the global supply and demand balances to determine what U.S. prices are likely to be; and the U.S. Secretary of Agriculture uses the report in establishing policy. The report includes projections of demand for a year ahead that usually rely on an extension of current trends of food consumption. USDA also projects long-run trends in consumption for the President's budget. The current practice is to extend current trends. This is a reasonable action in the absence of better information, but it has become increasingly troublesome as commodity demand-shifting changes in dietary preferences, as well as other factors, have been observed, sometimes during relatively short time frames.

Because the *WASDE* report projects commodity disappearance about 18 months into the future, U.S. food demand is largely assumed to be static. Eating habits are assumed to change slowly over time and changing diets are difficult to document even after the fact. For example, how much has per capita beef consumption risen as a result of low-carbohydrate diets? Access to grocery store scanner data and private survey data could provide some insight into this and other such questions. The disappearance data do not allow such tracking of demand factors.

Various problems are encountered when gathering data about food consumption. For example, lack of demand information limits the ability of analysts to track emerging trends in the meat sector. The export market and away-from-home consumption are gaining importance in the meat sector. These segments likely have different demand elasticities from the retail sector, and the price response from a change in supply may be different from those measured when the retail sector was the predominant source of meat consumption. Thus, it is important to have data on where food is consumed—something that is currently lacking for such analyses.

FOOD CONSUMPTION DATA AND THE EVALUATION OF FOOD ASSISTANCE PROGRAMS

U.S. food and nutrition assistance programs are intended to provide a nutritional safety net. The aim of the programs is to eliminate hunger and guarantee freedom from want by eliminating the worry about lack of food. Food and nutrition programs also promote healthy life-styles through nutrition education and by offering guides to dietary choices. Expenditures on food and nutrition assistance programs in fiscal year 2003 totaled $41 billion. Several participants noted that such a large expenditure warrants evaluation and that enhancements to the current data infrastructure could help to determine whether existing programs accomplish their purpose, to identify new nutritional challenges, and to aid the design of policy.

Programs: Administrative Information Needs

Various administrative data are needed to evaluate food and nutrition assistance programs. Jay Hirschman from the Office of Analysis, Nutrition, and Evaluation of USDA's Food and Nutrition Service gave as an example data needed to assess federal meal programs, such as school lunches and breakfasts. First, USDA must know what food was purchased by the institution, then how the food was offered (such as a meal, a la carte, through vending machines, or in the school store). Then it must know what food was consumed at the target meal and what food was consumed over 24 hours. Many factors influence what students are offered and what they are actually eating. Food and production costs, production methods, and menu planning systems affect what an institution offers. The availability and price of competitive meals and plate waste are also important issues for which more information would be useful.

David Smallwood from ERS also spoke about the data needs of food programs and of vendors and suppliers. Analysts need to know the cost to acquire and deliver benefits for food programs. They must understand client access, participation rates, and how participants use their benefits—that is, what food is being purchased and consumed. Vendors and suppliers need to know how access and availability of foods and costs affect consumption. They must understand channels of distribution and the effect of food labeling. They need to know the effects of certain events on pricing.

Program Monitoring and Evaluation

Program monitoring and evaluation are essential to providing effective and beneficial food assistance programs. Hirschman gave as an example a description of the School Nutrition Dietary Assessment Study III. One of the charges to the study is to determine student dietary intakes and the effect of USDA meals on students' specific meal and total dietary intake at school and over 24 hours. The study will compare the nutrient intake of participants to the nutrient intake of nonparticipants who consume meals brought from home; nonparticipants who purchase meals a la carte from the school cafeteria; and nonparticipants who obtain food from other sources. It will also study the effect of school meal participation on a specific day of nutrient intakes of school children. For policy development and program management, food consumption data are best collected in conjunction with program variables that provide insight on action likely to lead to improvement. For example, USDA does not make policy on what children must eat at school, but it does make policy on what students are offered and served; on what price students of different income levels pay; on education and promotion provided by schools; and on the school eating environment.

Hirschman also stated that general surveys, such as NHANES, provide valuable descriptive information, but are less useful for program evaluation because they do not involve random assignment of the intervention (in this case, program participation) and a control group. Special surveys and evaluations can be designed to more specifically target program use and effects of program use and are therefore more useful and meaningful information for program evaluation. An ongoing program of special surveys and evaluations could assess alternatives to current programs or program components. There are many gaps in information on food consumption data and evaluation of food programs. For example, what is the longitudinal relationship between Food Stamp Program participation and hunger and dietary status? What are the effects of free fruit and vegetables in schools on overall dietary intake? What are the very long-term effects of prenatal participation in the WIC (Women, Infants, and Children) program? What is the most effective balance of food assistance, nutrition education and promotion, physical activity education promotion, and environmental change in order to simultaneously prevent both hunger and obesity?

Smallwood spoke about the major limitations of current food consumption research. Research dollars are limited. Currently available data are also relatively old. Public data take years to gather and release. Once data are available, a model and a framework are needed to appropriately analyze and understand them. Current public surveys, such as NHANES, used to gather only 1 day of food intake data, though NHANES began collecting 2 days of intake per person in 2002. While 1 day of food intake is useful for estimating the population's mean intake, at least 2 days of food intake data are needed for estimating the distribution of usual nutrient intake and assessing nutrient inadequacy in relation to the recommended levels.[2] There are special data needs at the consumer level. Food choices, diet quality, food prices, food expenditures, and food security all affect what people eat. Food consumption can affect obesity, cognitive development, and health status and well being. Socioeconomic and demographic data can help researchers understand the interplay between such variables as race, poverty, and food consumption.

Program participation and eligibility status are key pieces of data for researchers. It is often difficult to determine whether people participate or are eligible to participate in the major food and nutrition assistance programs. For example, according to Smallwood, there is an underreporting of 20 percent or more of Food Stamp Program participation on the Current Population Survey, and, according to Steven Haider of the Department of Economics at Michigan State University, the participation rate for those eligible for the Food Stamp Program is probably under 50 percent. Understanding why people decide to participate in food programs requires an accurate estimate of eligibility. Most datasets are weak on eligibility information, especially those with the best outcome information. Some participants noted that NHANES offers the best information on nutrition outcomes, but it does not have detailed questions to estimate eligibility.

Haider discussed food expenditures and their relationship to nutrition outcomes. Food expenditure information is absolutely central to understanding nutrition outcomes. Are low-quality diets associated with low food

---

[2]*Dietary Reference Intakes. Applications in Dietary Assessment.* Subcommittees on Interpretation and Uses of Dietary Reference Intakes and Upper Reference Levels of Nutrients, Standing Committee on the Scientific Evaluation of Dietary Reference Intakes, Food and Nutrition Board, Institute of Medicine. Washington, DC: National Academy Press, 2000.

expenditures? If not, is this due to lack of dietary information or other constraints? The CSFII collected information on food expenditures, but when CSFII was merged with NHANES, food expenditure data were no longer collected. NHANES is now the major dataset for understanding nutrition outcomes, but the missing food expenditure data limit a full understanding of the relationships described above. Haider said that even a single question on food expenditures added to NHANES would be very useful.

Data linkage could improve nutritional program evaluation. Jay Bhattacharya from Stanford University stated that nutritional outcomes are the result of a complex process in which individual decisions interact with constraints imposed by the environment, such as family, job, school, and state and federal policy. The ability of researchers to link data together is critically important to study this complex process. By linking existing data and developing new data, more can be learned about particular populations, such as low-income people or children, and about institutions, such as schools, food vendors, agricultural markets, and government agencies.

Given limited resources to conduct population surveys, Bhattacharya enumerated two strategies. First, overlapping datasets could be developed, none of which cover everybody or everything, and linked together statistically. Different surveys could be aimed at different populations, such as the poor (Survey of Program Dynamics) or food stamp recipients (National Food Stamp Program Survey). Different surveys have different data strengths. For example, NHANES provides a medical and laboratory examination, CSFII included household food expenditures, and the CE includes food and other household expenditures. Second, a comprehensive data collection effort could be undertaken. A comprehensive strategy could include a health assessment module like NHANES; it could be collected as frequently as the Current Population Survey (CPS); it could include a longitudinal component like the National Longitudinal Survey of Youth; it could oversample needy populations, as does the Survey of Program Dynamics; it could include economic data like the Health and Retirement Study; and it could include a household food expenditure model, as did the CSFII.

If little extra money for surveys is available, Bhattacharya suggested exploiting state and federal administrative databases and making bold data linkages possible, while guaranteeing confidentiality. If lots of extra money were available, there should be a shift to a comprehensive strategy and larger sample sizes and economic modules in order to fill some key data gaps.

FOOD SAFETY AND FOOD CONSUMPTION DATA

There are several agencies with food safety responsibilities in the federal government. Representatives of the Food Safety and Inspection Service (FSIS) of USDA, the Center for Food Safety and Applied Nutrition (CFSAN) of the Food and Drug Administration in DHHS, and the Office of Pesticide Programs (OPP) of EPA made presentations at the workshop. FSIS ensures that meat, poultry, and egg products shipped in commerce for human food are safe, wholesome, and accurately labeled. CFSAN is responsible for promoting and protecting public health by ensuring that the nation's food supply is safe, sanitary, wholesome, and honestly labeled. OPP is responsible for regulating the nature and amount of pesticide residues in food.

Food Safety and Inspection Service

Philip Derfler of FSIS said that his office needs data to help identify public health issues and trends, and to identify and react to public health risks. FSIS uses food consumption data to estimate benefits and costs in food safety and nutrition regulatory impact analysis; to perform risk assessments to assess exposure to meat and poultry products and how people obtain and use information about the safety and nutritional characteristics of products; and to develop and implement policies on labeling and related education programs. FSIS uses data from NHANES and the old CSFII and Diet and Health Knowledge Survey (DHKS), but there are gaps in the data. For example, perceptions of products of new technologies and some commodities are missing. The absence of recent and reliable data hampers the ability to demonstrate economic benefits of changes in nutrition, safe handling, and other labeling rules. DHKS data are currently 8 years old; the survey is no longer conducted. There are concerns about data quality when using these data because they are so out of date.

Available up-to-date data on food consumption of high-risk products by various populations would help FSIS better assess risks of food-borne illness from pathogens. Food production trends and marketing data could help FSIS assess products of new technology on the market and which population groups purchase such products. They would also help FSIS determine trends for allergen labeling requirements. FSIS could better evaluate breaches in safe food handling. Available up-to-date data on food consumption by various populations would help FSIS assess product recall situations. Regulatory activities would be more effective if FSIS had access to production and marketing trends data, data on diet and health knowledge, and data on health status. FSIS uses survey data on nutrient intake and label use to correlate intake of nutrients with the use of existing nutrition facts. Data on how people obtain and use information on characteristics of meat and poultry products would help suppliers understand the economic effects of new requirements and assess enhanced health benefits to consumers. Finally, FSIS could use up-to-date data on health status and food consumption to determine the adequacy of petitions for nutrient fortification of meat and poultry products.

Michael DiNovi from CFSAN reviewed his center's program needs. For its food and ingredient safety assessments, the center needs various exposure assessments before using new additives, amending uses of current ingredients, or issuing generally recognized as safe (GRAS) notifications. CFSAN also needs data for its contaminant risk assessments for both naturally occurring and human-made contaminants.

DiNovi said that CFSAN's Total Diet Study (TDS), which determines levels of various contaminants and nutrients in foods, is derived from the CSFII or NHANES (depending on the most recent source of information) and includes a seasonal collection of market baskets, a monitoring of trends, and a monitoring of new substances of health concerns. The TDS involves purchasing samples of food throughout the United States, preparing the foods as they would be consumed (table ready), and analyzing the foods to measure the levels of contaminants or nutrients of interest. Food samples are purchased by FDA personnel from supermarkets or grocery stores in selected cities, and the samples are sent to FDA laboratories for analysis. Dietary intakes of these analytes are then estimated for the U.S. population by multiplying the levels found in the TDS samples by the amounts of foods consumed based on surveys, such as the CSFII, conducted by the USDA (see http://www.cfsan.fda.gov/~comm/tds-hist.html). As previously noted, the CSFII was discontinued, and the data are aging rapidly. CFSAN's applied nutrition programs are using NHANES data for its Preventing Obesity Through Better Nutrition Project.

CFSAN's future needs include continuing access to multiday food intake surveys. DiNovi and other participants noted that if any extra food consumption research money were to become available, it should be spent on increasing the number of days of dietary recall. But Susan Krebs-Smith of NIH and Barry Popkin of the University of North Carolina stated in a later workshop discussion that increasing 2-day dietary recall by 1 or 2 days would not increase understanding of usual dietary intake. They said that a food propensity questionnaire provides more information on usual intake than 3 or 4 days of dietary intake recalls could provide. CFSAN would also like 24-hour dietary supplement recall.

Office of Pesticide Programs

David Miller of OPP gave a presentation on his office's use of the CSFII. OPP estimates dietary exposure to pesticides based on two separate data sources: the CSFII 1994-1996/1998 and the amount of pesticide in and on food, which comes from field trial data, monitoring data, and market basket survey data. The CSFII is used because it is a nationally representative and statistically based survey that gathered data on the intakes of individuals all seasons of the year and all days of the week. It included a large number of individuals, and populations of interest were oversampled. Finally, CSFII was a high-quality survey, including in-person 24-hour dietary recalls, and it had a high response rate (roughly 75 percent for the 2-day dietary recall). It also included extensive ancillary and demographic data. Miller noted that OPP will rely more heavily on NHANES data in the future now that the CSFII has been discontinued. OPP looks forward to using the NHANES food propensity questionnaire. Miller is concerned that NHANES will not offer a large enough sample size to study small populations in depth.

Food Safety—Data Users

Neal Hooker from the Department of Agricultural, Environmental, and Development Economics at the Ohio State University provided a data user's perspective of food safety. The research community has been working for years to fill some data gaps. The Cost of Illness Calculator, available through USDA's Economic Research Service (ERS), is a useful and customizable tool for estimating the cost of food-borne illnesses. CDC provides another very useful resource—FoodNet—which identifies emerging food-borne infections. FDA's Operation and Administrative System for Import Support (OASIS) is the most important database for tracking potential problems with imports, but these data are not collected randomly, and the sample size is small relative to the volume of imports. Recall data (when products are recalled from the market due to food safety hazards) help highlight emerging problems. They could be even more useful if they were linked to other analyses, like food processing plant data or census data.

Food labeling is one of the most dramatic recent changes to diet and health knowledge in the United States. Direct surveys, focus groups, or experiments can be used to assess the effect of qualified health claims and other information on food labels.

FOOD CONSUMPTION DATA AND HEALTH

Eric Hentges from the USDA's Center for Nutrition Policy and Promotion (CNPP) began by saying that good data lead to good policy. Better data would lead to better policy. It is imperative to gather data on diet knowledge and attitudes. By gathering this type of data, informed decisions can be made, for example, about what kinds of policies will lead people to eat more fruits and vegetables. Both the food guide pyramid and food labeling methods are built on consumption data. Quality data are essential to policy and intervention programs.

Hentges noted that baseline and longitudinal data on such markers as weight and cholesterol are needed to better understand problems like the obesity epidemic. Hundreds of millions of dollars are spent on research trials, but little or no money is spent on gathering data from these trials, according to Hentges.

Currently, many food assistance programs run on performance-based budgets. If an agency cannot prove that its program succeeds at its mission, the program can lose some or all of its funding. If agencies do not have foundational, baseline data, they cannot possibly begin to show that their programs have had an effect on the target population. This is yet another reason why good consumption data are needed, Hentges noted.

Food Consumption Data for Cancer and Other Disease Research

Susan Krebs-Smith from the Risk Factor Monitoring and Methods Branch of the NIH's National Cancer Institute gave a presentation on her organization's interest in food consumption data. The mission of the Risk Factor Monitoring and Methods Branch is to contribute to reducing cancer in the U.S. population by serving as a critical link between research on the causes and origins of cancer risk factors and targeted interventions for prevention. Many of the risk factors that the agency studies are risk factors not only for cancer, but for other chronic diseases as well. The agency develops and improves methods to assess such factors and provides data to assist in formulating public policies.

The agency's areas of research include tobacco use; diet; weight, height, and related measures; physical activity; genetics, family history, and individualized risk assessment; sun exposure; and pharmaceutical use. The agency also studies many diet-related topics. It conducts surveillance, such as the prevalence of consuming a certain number of fruits and vegetables daily. It also monitors health objectives, health policy, program evaluation, and health disparities. Krebs-Smith noted that the national surveys do not have a sufficient sample size to study many health disparities. It is difficult to determine differences among groups in dietary patterns with a sample size of only about 5,000 persons per year.

Food consumption survey data are the most direct measure of dietary food intake. Food consumption data can measure nutrients, foods as eaten, food guide pyramid servings, and agricultural commodities. Food consumption data can provide details about timing, patterns, and combinations of foods.

National food consumption survey data have limitations. The annual sample size in current surveys is inadequate for many policy-relevant analyses, especially since the CSFII was discontinued. Discontinuing the CSFII was a loss in terms of national dietary data. Measurement error is a problem with 24-hour dietary recalls. There needs to be a way to assess usual intake because there are great within-person variations. Usual intake of rare foods is impossible to examine with only one 24-hour recall. Dietary supplement intakes are not as well quantified as foods, so they are difficult to incorporate into nutrient intake measurements. The final limitation Krebs-Smith noted was food cost information. It would be helpful to have cost information tied with diets.

Food supply data fill some gaps in understanding the American diet that cannot be filled with food consumption survey data. Food supply data address aggregate consumption and provide researchers with upper bounds on food intake. Also, a consistent methodology has been applied to the food supply data over time so it is possible to study trends over time. This is not the case for food consumption data. Food supply data can also reveal the agricultural implications of eating according to the food guide pyramid recommendations.

Food supply data have some limitations. They cannot be related to health disparities, the relation of diets to other health factors cannot be studied, and food supply data do not offer details about how foods are consumed.

Krebs-Smith noted that NIH is also interested in filling gaps in assessment of diet, weight, and physical activity. To that end, the agency has developed two new modules for use in NHANES. Although 24-hour recalls provide important food consumption information, in order to attain a good estimate of the distribution of intake of specific food and nutrients it is imperative to assess usual intake. With this in mind, the agency developed a food propensity questionnaire that has recently been added to NHANES. The agency also saw the need for an objective measure of physical activity, so the agency has supported the addition of a physical activity monitor to NHANES. The agency is also working to add some questions on personal weight-loss efforts, health practitioner advice on weight loss, and weight loss history.

Subpopulations are difficult to monitor, even with a survey as large and inclusive as NHANES. Krebs-Smith mentioned the idea of a community HANES, which would include scaled-down mobile examination centers that would be able to capture variables such as diet, physical activity, anthropometrics, and biomarkers in one visit. Populations defined by race and ethnicity or by various geographic areas could be studied in depth. Because these populations would be geographically concentrated, collection of community-level variables would be possible. Census-tract level information on neighborhoods—such as the availability of food sources and access to walking areas—could easily be collected.

Barry Popkin of the Department of Nutrition at the University of North Carolina at Chapel Hill gave a presentation focusing on the importance of data linkage and a critique of NHANES for understanding diet and health links. The CSFII and other past USDA surveys had no state or local identifiers or ability to link to price and other contextual data. This lack makes it very difficult to study such issues as state or county-level WIC programs or to evaluate school lunch programs. One of the challenges for NHANES is to be disaggregated to a level at which the richness of other datasets can be fully used. Geocoding could be used to achieve this. In order to study the determinants of dietary behavior, it is necessary to be able to link individual and household data to food price data at the smallest geographic units possible. A vast array of other contextual issues need to be studied to understand how the broader environment affects food choice. Many researchers would like linkage to the actual address of interviewees, and there are ways to do so that would protect the privacy and confidentiality of human subjects.

Popkin then spoke about issues relating to NHANES. Changes in NHANES coding affect trend, program, and policy analysis. He also noted that NHANES never did bridging studies to understand the effects of changes in methods used to collect data between the key phases of diet collection design in the 1980s and the 1990s.

Popkin noted that there are no national datasets in the U.S. that collect dietary data longitudinally. USDA has funded some hunger questions on the Early Childhood Longitudinal Survey (ECLS), but there are no questions relating to diet. The National Institute of Child Health and Human Development, which cooperates with the Department of Education to conduct the ECLS, has proposed a new birth cohort study, the National Children's Study. It would examine the effects of environmental influences on the health and development of more than 100,000 children across the United States, following them from before birth until age 21. The goal of the study is to improve the health and well-being of children (see http://nationalchildrensstudy.gov/about/overview.cfm). Unfortunately, the study does not currently include plans for collecting dietary data. This is a missed opportunity, Popkin said.

PROPRIETARY DATA FOR POLICY QUESTIONS

Abebayehu Tegene began this session by noting that public sources of data may not be sufficient for research or for policy analysis, for which timely data are essential. The private sector could provide some data files of use to researchers and policy makers. The private sector's infrastructure allows it to conduct focused surveys very quickly and provide customized reports. Private-sector data may not be best for answering questions of diet and health, but they can help look at how market forces influence diet and health. A third provider of proprietary data, IRI (Information Resources, Inc.), was invited to speak at the workshop, but was unable to send anyone. Food consumption data include both dietary intake data generally gathered through dietary recalls and food propensity (or frequency) questionnaires.

ACNielsen

John Green, vice president of industry strategy at ACNielsen, presented an overview of his company's work relating to food consumption and purchases. ACNielsen gathers information about both retailers and consumers. Virtually all retailers except Wal-Mart send ACNielsen price and item information, on a weekly basis. ACNielson edits and processes the data. The company works mostly for manufacturers, retailers, food brokers, and wholesalers. The company does occasional work for the federal government on an ad hoc basis.

ACNielsen also does survey work through its household panels. Through this survey work, attitudes and behaviors can be linked. Household panel participants are given a scanner that they keep at home. Each time a household member shops, he or she uses the scanner to record each item purchased. For items without a universal product code (UPC), such as some fresh meats and produce, respondents are given a vocabulary code book to identify the item, and they are asked for the product's weight. Once a week participants send their information to ACNielsen by telephone lines. Information gathered by ACNielsen includes what store was shopped in and the age and sex of the shoppers. Currently, ACNielsen processes 61,000 households every week, and the company is planning a major expansion of the household panels. One of the limitations of these data, however, is that the company has difficulty recruiting certain kinds of households, such as those of minorities, low-income families, and mobile singles.

NPD Group

Cindy Beres, operations manager-Foodworld of the NPD Group, presented an overview of her company's work relating to food. The company's Consumer Report on Eating Share Trends (CREST) tracks consumer purchases of prepared meals and snacks from commercial restaurants. CREST is a daily online survey of about 3,000 adults and 500 teens. Behavioral and attitudinal survey questions are included. The CREST survey captures what participants ate yesterday, where they purchased it, where they ate it, who they were with, and how much money they spent.

The NPD Group's National Eating Trends (NET) tracks food and beverage preparation and consumption habits, including end dishes, ingredients, additives, and cooking aids. NET has been continuous since 1980; it includes 14 consecutive daily food and beverage diaries, which are returned daily. About 60 households begin 14-day diaries every Monday. Data are accessible 3 months after the close of the quarterly data collection period. NPD is currently conducting a supplement of 500 Hispanic households. The NET database variables include description of the food (kind, flavor, type), how it was served (topping, main dish), the brand name, where it was obtained, how it was prepared, the ingredients, and who consumed the food.

After collecting the 14 days of food intake, NPD collects the following information:

- diet status
- diet type
- medical conditions
- height and weight (self-reported)
- vitamin and mineral supplement usage
- vegetarian or not
- exercise (number of days per week, type, and history)
- nutritional attitudes (concerns about sugar, carbohydrates, taste, trans fat, etc.).

NET is similar to national government surveys in that it samples about 5,000 individuals annually, collects similar food details and brand names, and has similar reporting capabilities. However, NET is different in other ways: the primary food preparer reports for all household members; it features a longitudinal design (the 14 days of data collection); and it offers continuous data collection and quarterly data releases. NET's food journal eliminates interviewer bias, but there is no opportunity to prompt for forgotten foods.

Data from the NPD Group's nutrient intake database on associated average serving size and nutrient composition are combined with data from NET on eating frequencies to estimate an individual's intake of macro and selected micro nutrients. The NPD Group mapped NET eating frequencies to the CSFII's average serving sizes by gender and age and the CSFII Survey Nutrient Database to create the Nutrient Intake Database. This database is currently available for 1998-2003.

Beres ended her discussion by mentioning two other NPD Group services—food safety monitor and dieting monitor. The food safety monitor regularly measures consumers' level of concern about various food safety issues. The dieting monitor regularly measures awareness and participation of popular diets.

Scanner Data Applications

Helen Jensen of Iowa State University began by describing the current consumer market. Incomes have been rising and labor markets are changing. Demographics are changing; the country is becoming more ethnically diverse. Scientific discoveries of food consumption and health links are making people look more closely at specific foods. New food technologies create challenges for traditional food groupings. For example, orange juice can now be fortified with calcium, so people may now be getting calcium from an unexpected source. The food industry's influence on dietary choices is also an emerging area of study. Because of all of these new issues, food product detail is often required to support research and decision making.

Household scanner data provide a good amount of product detail and space and time purchase information. They also allow purchases to be matched with demographics. However, there are some problems with data quality. Products with standard UPC codes are easily picked up and recorded with scanner data, but such products as fruits, vegetables, and some cheeses and meats lack standardized UPC codes. The level of product detail can also be a disadvantage because researchers must narrowly define their categories and then seek out the appropriate products from long lists.

Scanner data can be used in evaluation and policy analysis. Scanner data can shed light on market participation: for example, what percentage of consumers is purchasing this product? Researchers and policy makers can also use scanner data to estimate demand parameters for a single product or a group of products. Scanner data can be used to determine both market and nonmarket evaluation: that is, the data can reveal both the market price and the value that consumers place on attributes in products. Scanner data include information on expenditures and expenditure sales, which can be particularly useful if a household receives food stamps or WIC benefits. Policy makers can then study how these households spend their money on food. Scanner data can also provide insight into infrequently purchased products or products only purchased by a few households. Scanner data can inform research and policy issues related to the introduction, adoption, purchase patterns, and demographic factors of new products.

Jensen also offered some criticisms of the ACNielsen HomeScan Panel. The panel members have a higher income and smaller household size than the general population, and they are more likely to be married, more likely to be white, and less likely to be Hispanic. When working with scanner data, it is important to think about whether the sample is representative and if low-income and minority populations are included in sufficient numbers. It is also important to think about whether store purchases are representative: purchases from convenience stores and other small stores may not be included. It is also important to know if food assistance program purchases can be identified.

POSSIBLE DATA IMPROVEMENTS AND DATA LINKAGES

During this session, four members of the Panel on Enhancing the Data Infrastructure in Support of Food and Nutrition Programs, Research, and Decision Making discussed possible data improvements or data linkages. Ronette Briefel enumerated six areas in which USDA could apply additional funding. First, USDA could augment existing data collections: for example, household assets and expenditure data could be added to NHANES, though it would be important not to overload that survey. Second, USDA could work to more thoroughly analyze existing data collections: there is a lot of untapped information in NHANES 1999-2000 and NHANES III (1988-1994). Third, work could be done to enhance methodological research areas, like dietary knowledge and attitudes, which lags 5-10 years in comparison with other areas, such as dietary intake methodology and physical activity assessment. Fourth, the advantages and disadvantages of cross-sectional and longitudinal data could be studied. What population groups and issues should be studied cross-sectionally or longitudinally? Fifth, information could be gathered about the school nutrition environment, such as what children are offered at meals, what foods are available in vending machines or other sources, what foods are purchased, and where children consume meals. The Youth Risk Behavior Survey or the Behavioral Risk Factor Surveillance Survey could be useful monitoring tools. Finally, datasets could be linked, and they could be developed using common definitions and survey questions on dietary behavior, attitudes, and sociodemographic factors. Briefel added that the discontinuation of the CSFII resulted in a loss of a sample of 5,000 persons. This loss of sample size limits researchers' abilities to add new modules to food consumption surveys and to study subpopulations.

Laurian Unnevehr raised the issue of linking macrodata with microdata: for example, food disappearance data could be linked to the NHANES to see whether or not one predicts the other. If a prediction is found, this could help researchers understand how short-term microlevel data collection could predict outcomes for national agriculture trends. Household scanner data, including surveys regarding concerns and attitudes, could be linked to national sales trends to see if attitudes and beliefs really affect people's purchases. Assessing the strength of such linkages would demonstrate whether one data source could be substituted for another. She emphasized that the loss of consumption data linked to both economic variables and diet/health/knowledge information—because of the demise of CSFII—is an important gap in the ability to answer policy questions. Some way to link economic data with food consumption data could also be considered, as well as a way to link health knowledge with actual consumption behavior. Alternative public investments in a survey that links food purchases with household economics, knowledge, and behavior need to be evaluated, she said. It is possible that a partitioned NHANES could meet these needs, but if not, then some new survey including all three kinds of information, and possibly with more limited consumption detail, could be considered.

Alan Kristal raised the issue of linking NHANES data to Social Security information in order to learn more about respondents' Medicare participation and income information. He discussed "conceptual linkage," meaning to develop sets of items that allow certain kinds of parallel analyses across different kinds of surveys. This approach would take a bit of scientific effort: diet knowledge, attitude, and behavior are difficult to measure, and there is currently no agreement on their definitions. Kristal also suggested consideration of overlapping sampling units across some of the large surveys: for example, have NHANES and another large government survey overlap in the same geographic area.

William Eddy of Carnegie Mellon University urged consideration of increasing the sample size and number of questions related to diet and demographics in NHANES. He also suggested consideration of interagency cooperation regarding food consumption data.

SUMMARY

The Workshop on Enhancing the Data Infrastructure in Support of Food and Nutrition Programs, Research, and Decision Making covered various topics related to food consumption over the course of its one-and-a-half day meeting. The workshop began with descriptions of key datasets, such as the NHANES, CE, and proprietary datasets. Representatives from various government agencies spoke about the specific food data needs of their offices. Researchers working outside of the federal government gave presentations that voiced their concerns about food consumption data and possible ways to improve the data infrastructure. The workshop provided an opportunity for people from government, private industry, and academia to come together and share concerns and ideas about data on food consumption and expenditures.

One presentation, by Jay Bhattacharya, gave two alternatives for how the data infrastructure could be improved. Overlapping datasets could be developed, none of which cover everybody or everything. Different surveys could be aimed at different populations, such as the poor or food stamp recipients. Or a comprehensive data collection effort could be undertaken that would include a health assessment module that is collected frequently, oversamples needy populations, includes economic data, and includes a household food expenditure model. Many participants expressed concern about the discontinuation of the CSFII—the loss of the 5,000-person sample and the loss of diet and health knowledge questions and questions on food expenditures.

The Panel on Enhancing the Data Infrastructure in Support of Food and Nutrition Programs, Research, and Decision Making considered each of these topics and others for its final report. The panel considered priority areas for new questions to surveys such as the NHANES that could fill gaps in knowledge about how people make food consumption and expenditure decisions. It also considered how alternative data sources, such as those from proprietary firms, could be used to fill gaps.

Appendix B

Workshop Agenda

**ENHANCING THE DATA INFRASTRUCTURE
IN SUPPORT OF FOOD AND NUTRITION PROGRAMS,
RESEARCH, AND DECISION MAKING**

The Melrose Hotel
2430 Pennsylvania Avenue, NW
Washington, DC

Workshop Agenda
May 27-28, 2004

Thursday, May 27

10:30 **Welcome and Introductions**
John Karl Scholz, *Chair, University of Wisconsin–Madison*
Constance Citro, *Director, Committee on National Statistics*
Susan Offutt, *Administrator, Economic Research Service, U.S. Department of Agriculture (USDA)*
Clifford Johnson, *Director, NHANES Program, National Center for Health Statistics, U.S. Department of Health and Human Services (DHHS)*

11:00 **Session 1**
Overview of Food Consumption, Expenditures, and Sales Datasets

Session Chair: John Karl Scholz, *University of Wisconsin–Madison*
James Blaylock, *Economic Research Service, USDA*
Steve Henderson and Sioux Groves, *Bureau of Labor Statistics, U.S. Department of Labor*
Clifford Johnson, *National Center for Health Statistics, DHHS*
Abebayehu Tegene, *Economic Research Service, USDA*

12:00 Lunch

1:00 **Session 2**
Food Marketing and Promotion and Food Market Analysis

Session Chair: F. Jay Breidt, *Colorado State University*
Donald Hinman, *Agricultural Marketing Service, USDA*
Gerald Bange, *World Agricultural Outlook Board, USDA*

2:00 **Session 3**
Food Consumption Data and the Evaluation of Food Assistance Programs

Session Chair: John Karl Scholz, *University of Wisconsin–Madison*
Jay Hirschman, *Food and Nutrition Service, USDA*
David Smallwood, *Economic Research Service, USDA*
Jay Bhattacharya, *Stanford University*
Steven Haider, *Michigan State University*

3:15 Break

3:30 **Session 4**
Food Safety and Food Consumption Data

Session Chair: Laurian Unnevehr, *University of Illinois, Urbana-Champaign*
Phil Derfler, Robert Post, Ron Meekhof, *Food Safety and Inspection Service, USDA*

Michael DiNovi, *Office of Food Additive Safety, Food and Drug Administration, DHHS*
David Miller, *Office of Pesticide Programs, Environmental Protection Agency*
Neal Hooker, *Ohio State University*

4:45   Open Discussion

5:00   Adjourn

Friday, May 28

8:30   Breakfast

9:00   Session 6
Food Consumption Data, Diet, and Health

Session Chair: Alan Kristal, *Fred Hutchinson Cancer Research Center, University of Washington*
Eric Hentges, *Center for Nutrition Policy and Promotion, USDA*
Susan Krebs-Smith, *National Cancer Institute, National Institutes of Health*
Barry Popkin, *University of North Carolina, Chapel Hill*

10:15   Break

10:30   Session 7
The Use of Scanner Data and Other Proprietary Sources of Data to Address Policy Questions: Panel Discussion

Session Chair: William Eddy, *Carnegie Mellon University*
Abebayehu Tegene, *Economic Research Service, USDA*
John Green, *Vice President of Industry Strategy, ACNielsen*
Cindy Beres, *Operations Manager, Foodworld, The NPD Group*
Gary Thompson, *University of Arizona*
Helen Jensen, *Iowa State University*

11:45 **Session 8**
Possible Data Improvements or Data Linkages: Panel Discussion

Session Chair: John Karl Scholz, *University of Wisconsin–Madison*
Ronette Briefel, *Mathematica Policy Research, Inc.*
Laurian Unnevehr, *University of Illinois, Urbana-Champaign*
Alan Kristal, *Fred Hutchinson Cancer Research Center, University of Washington*
William Eddy, *Carnegie Mellon University*

12:30 **Workshop Adjourns**

Appendix C

Workshop Participants

PRESENTERS

Gerald Bange, World Agricultural Outlook Board, U.S. Department of Agriculture
Cindy Beres, Operations Manager, Foodworld, The NPD Group
Jay Bhattacharya, Center for Primary Care and Outcomes Research, Stanford University
James Blaylock, Economic Research Service, U.S. Department of Agriculture
F. Jay Breidt, Department of Statistics, Colorado State University
Ronette Briefel, Mathematica Policy Research, Inc.
Constance Citro, Committee on National Statistics, The National Academies
Phil Derfler, Food Safety and Inspection Service, U.S. Department of Agriculture
Michael DiNovi, Office of Food Additive Safety, Food and Drug Administration, U.S. Department of Health and Human Services
William Eddy, Department of Statistics, Carnegie Mellon University
Andrew Gelman, Department of Statistics, Columbia University
John Green, Vice President of Industry Strategy, ACNielsen
Sioux Groves, Bureau of Labor Statistics, U.S. Department of Labor

---

NOTE: All affiliations are as of the time of the workshop.

Steven Haider, Department of Economics, Michigan State University
Steve Henderson, Bureau of Labor Statistics, U.S. Department of Labor
Eric Hentges, Center for Nutrition Policy and Promotion, U.S. Department of Agriculture
Donald Hinman, Agricultural Marketing Service, U.S. Department of Agriculture
Jay Hirschman, Food and Nutrition Service, U.S. Department of Agriculture
Neal Hooker, Department of Agricultural, Environmental, and Development Economics, Ohio State University
Susan Krebs-Smith, National Cancer Institute, National Institutes of Health
Helen Jensen, Department of Economics, Iowa State University
Clifford Johnson, National Center for Health Statistics, U.S. Department of Health and Human Services
Alan Kristal, Fred Hutchinson Cancer Research Center, University of Washington
Ron Meekhof, Food Safety and Inspection Service, U.S. Department of Agriculture
David Miller, Office of Pesticides Programs, Environmental Protection Agency
Susan Offutt, Economic Research Service, U.S. Department of Agriculture
Barry Popkin, Department of Nutrition, University of North Carolina, Chapel Hill
Robert Post, Food Safety and Inspection Service, U.S. Department of Agriculture
John Karl Scholz, Department of Economics, University of Wisconsin–Madison
David Smallwood, Economic Research Service, U.S. Department of Agriculture
Abebayehu Tegene, Economic Research Service, U.S. Department of Agriculture
Laurian Unnevehr, Department of Agricultural and Consumer Economics, University of Illinois at Urbana-Champaign

OTHER PARTICIPANTS

Nicole Ballenger, Economic Research Service, U.S. Department of Agriculture
Peter Basiotis, Center for Nutrition Policy and Promotion, U.S. Department of Agriculture
Mary Brandt, Center for Food Safety and Applied Nutrition, Food and Drug Administration, U.S. Department of Health and Human Services
Andi Carlson, Center for Nutrition Policy and Promotion, U.S. Department of Agriculture
Steven Carlson, Food and Nutrition Service, U.S. Department of Agriculture
Gary Crisafulli, ACNielsen
Ken Dalton, Bureau of Labor Statistics, U.S. Department of Labor
Joe Derochowski, Director of Business Development, National Eating Trends, The NPD Group
William Dietz, National Center for Chronic Disease Prevention and Health Promotion, Centers for Disease Control and Prevention, U.S. Department of Health and Human Services
Johanna Dwyer, Department of Society, Human Development, and Health, Harvard University
Kenneth Falci, Center for Food Safety and Applied Nutrition, Food and Drug Administration, U.S. Department of Health and Human Services
Barbara Fraumeni, Bureau of Economic Analysis, U.S. Department of Commerce
Debbie Gann, Research Director, Food Marketing Institute
Shirley Gerrior, Center for Nutrition Policy and Promotion, U.S. Department of Agriculture
Nancy Gordon, Associate Director for Demographic Programs, Census Bureau, U.S. Department of Commerce
Mary Hager, Senior Manager, Regulatory Affairs, The American Dietetic Association
Colien Hefferan, Cooperative State Research, Education, and Extension Service, U.S. Department of Agriculture
Teresa Hicks, Demographic Surveys Division, Census Bureau, U.S. Department of Commerce

Greg Key, Consumption Branch, Bureau of Economic Analysis, U.S. Department of Commerce
Betsey Kuhn, Economic Research Service, U.S. Department of Agriculture
Patricia McKinney, Food and Nutrition Service, U.S. Department of Agriculture
Steven Landefeld, Director, Bureau of Economic Analysis, U.S. Department of Commerce
Michael LeBlanc, Economic Research Service, U.S. Department of Agriculture
Biing-Hwan Lin, Economic Research Service, U.S. Department of Agriculture
Cristina McLaughlin, Center for Food Safety and Applied Nutrition Division of Market Studies, Food and Drug Administration, U.S. Department of Health and Human Services
Alanna Moshfegh, Agricultural Research Service, U.S. Department of Agriculture
Linda Myers, Center for Food and Nutrition Policy, Virginia Polytechnic Institute and State University
Kathy Radimer, National Center for Health Statistics, Centers for Disease Control and Prevention
Maria Reed, Demographic Surveys Division, Census Bureau, U.S. Department of Commerce
Susan Schechter, U.S. Office of Management and Budget
Arnie Schwartz, National Eating Trends, The NPD Group
Edward Sondik, Director, National Center for Health Statistics, Centers for Disease Control and Prevention, U.S. Department of Health and Human Services
Carol Spease, Center for Food Safety and Applied Nutrition, Food and Drug Administration, U.S. Department of Health and Human Services
Joseph Spence, Agricultural Research Service, U.S. Department of Agriculture
Amy Subar, National Cancer Institute, National Institutes of Health
Lorraine Thaden, Agricultural Research Service, U.S. Department of Agriculture
Jay Variyam, Economic Research Service, U.S. Department of Agriculture
Willis Wells, Cooperative State Research, Education, and Extension Service, U.S. Department of Agriculture
Jerry West, Early Childhood Longitudinal Study, National Center for Education Statistics, U.S. Department of Education

PANEL MEMBERS

F. Jay Breidt, Department of Statistics, Colorado University
Ronette Briefel, Mathematica Policy Research Inc., Washington, DC
William Eddy, Department of Statistics, Carnegie Mellon University
Andrew Gelman, Department of Statistics, Columbia University
Alan Kristal, Fred Hutchinson Cancer Research Center, University of Washington
Barry Popkin, Department of Nutrition, University of North Carolina, Chapel Hill
John Karl Scholz *(chair)*, Department of Economics, University of Wisconsin–Madison
Laurian Unnevehr, Department of Agriculture and Consumer Economics, University of Illinois at Urbana-Champaign

COMMITTEE ON NATIONAL STATISTICS STAFF

Constance Citro, *Director*
Michele Ver Ploeg, *Study Director*
Jamie Casey, *Research Associate*
Tanya Lee, *Project Assistant*
Jerusha Nelson Peterman, *Intern*

Appendix D

Biographical Sketches of Panel Members and Staff

John Karl Scholz (*Chair*) is a professor of economics at the University of Wisconsin–Madison. Previously, he was the deputy assistant secretary for tax analysis at the U.S. Department of the Treasury and senior staff economist at the Council of Economic Advisers. He has written extensively on the earned income tax credit and low-wage labor markets. He also writes on public policy and household saving, charitable contributions, and bankruptcy laws. He is a research associate at the National Bureau of Economic Research and was director of the Institute for Research on Poverty at the University of Wisconsin–Madison. He received a Ph.D. in economics from Stanford University.

F. Jay Breidt is professor and director of graduate education for the Department of Statistics at Colorado State University (CSU). Previously, he was on the faculty in the Department of Statistics at Iowa State University and a member of the survey section of the statistical laboratory, which had as a major focus design and estimation for large-scale environmental surveys, particularly the USDA's National Resources Inventory. His research interests include time series, environmental monitoring, and survey sampling. He is an associate editor of the *Journal of the American Statistical Association* and the *Journal of Forecasting*, and he currently chairs the American Statistical Association's Committee on Energy Statistics. With colleagues at CSU, he was recently awarded an EPA STAR grant for space-time aquatic resources modeling. He received a Ph.D. in statistics from Colorado State University.

Ronette Briefel is a senior fellow at Mathematica Policy Research, Inc. Her research interests include national nutrition policy; survey research on the dietary food security, nutritional, and health status of the U.S. population; and dietary intake methodology. She has analyzed National Health and Nutrition Examination Survey (NHANES) data on the dietary intake and nutritional status of low-income populations, including pregnant women and children participating in WIC and has conducted national program evaluations of the Summer Food Service Program and the Emergency Food Assistance Program. She earned a B.S. in nutrition from Pennsylvania State University and an M.P.H. in maternal and child health administration and a Dr.P.H. in epidemiology from the University of Pittsburgh's Graduate School of Public Health.

Jamie Casey (*Research Associate*) is a member of the staff of the Committee on National Statistics. She has worked on projects studying the Special Supplemental Nutrition Program for Women, Infants, and Children (WIC) eligibility, the State Children's Health Insurance Program (SCHIP), the role of Institutional Review Boards in social and behavioral research, and racial and ethnic disparities in health care. Previously, she worked for the National Center for Health Statistics. She received a B.A. degree in psychology from Goucher College.

Constance F. Citro (*Staff Director*) is director of the Committee on National Statistics. She is a former vice president and deputy director of Mathematica Policy Research, Inc., and was an American Statistical Association/National Science Foundation research fellow at the U.S. Census Bureau. She has served as study director for numerous projects, including the Panel to Review the 2000 Census, the Panel on Estimates of Poverty for Small Geographic Areas, the Panel on Poverty and Family Assistance, the Panel to Evaluate the Survey of Income and Program Participation, the Panel to Evaluate Microsimulation Models for Social Welfare Programs, and the Panel on Decennial Census Methodology. Her research has focused on the quality and accessibility of large, complex microdata files, as well as analysis related to income and poverty measurement. She is a fellow of the American Statistical Association. She received a B.A. degree from the University of Rochester and M.A. and Ph.D. degrees in political science from Yale University.

William F. Eddy is professor of statistics at Carnegie Mellon University, and he also holds appointments in the School of Computer Science and the Department of Biological Sciences. He is an elected fellow of the American Statistical Association, the Institute of Mathematical Statistics, and the American Association for the Advancement of Science, and is an elected member of the International Statistical Institute. He is the chair of the Committee on National Statistics of the National Academies and was previously chair of the National Academies' Committee on Applied and Theoretical Statistics. He received a Ph.D. degree in statistics from Yale University.

Andrew Gelman is a professor in the Department of Statistics and the Department of Political Science at Columbia University. He is the author of *Bayesian Data Analysis* and *Teaching Statistics: A Bag of Tricks*, and more than 100 research articles. His research interests include Bayesian methods, statistical graphics and computation, sample surveys, and applications in public health and policy. His research has won the Heinz Eulau Award from the American Political Science Association and the Outstanding Statistical Application Award from the American Statistical Association. He is an elected fellow of the American Statistical Association and the Institute of Mathematical Statistics.

Alan R. Kristal is a professor in the Fred Hutchinson Cancer Research Center at the University of Washington. His primary research interest is in nutritional epidemiology, including the etiologic relationships between diet and cancer and the implementation and evaluation of public health nutrition interventions. Current projects include studies on (1) diet, dietary supplements, and prostate cancer risk; (2) diet and neoplastic progression of Barrett's esophagus; (3) low-fat diet and breast cancer survival; and (4) dietary supplement use and cancer risk. He is a member of several medical associations and societies, including the American Association for Cancer Research, the American Society for Clinical Nutrition, the American Society for Nutritional Sciences, and the Society for Epidemiologic Research. A recent article, "Serum selenium levels in relation to markers of neoplastic progression among persons with Barrett's esophagus," was featured in the May 2003 issue of the *Journal of the National Cancer Institute*.

Earl S. Pollack (*Study Director*) is a member of the staff of the Committee on National Statistics. Previously, he was chief of biometry at the National Cancer Institute and director of the Division of Biometry and Epidemiology at the National Institute of Mental Health. More recently, he was a research professor at the Biostatistics Center at George Washington University and served as statistician for the Center to Protect Workers Rights, the construction research arm of the AFL/CIO. His interests are in chronic disease epidemiology and in the analysis of observational data from large health and medical databases. He is a fellow of the American Statistical Association, the American College of Epidemiology, and the American Public Health Association. He received B.S. and M.A. degrees in statistics from the University of Minnesota and an Sc.D. in biostatistics from Harvard University.

Barry M. Popkin is a professor of nutrition at the University of North Carolina. His primary focus is on the nutrition transition around the world, particularly the demographic and economic determinants of diet, activity, and body composition trends, through the use of longitudinal analysis techniques. Popkin directs longitudinal surveys in China and Russia; his long-term Russian Longitudinal Monitoring Survey provides data comparable to those from the China Health and Nutrition Survey and is the official monitor of Russian economic reforms and their demographic, health, and nutritional effects. He is also involved in longitudinal research in South Africa and the Philippines, and is working with researchers in several other countries. He is a consultant to the World Bank, the U.N. Coordinating Committee on Nutrition, the Micronutrient Initiative, and the U.N. Children's Fund. In 1998 he received the Kellogg Prize for Outstanding Research in International Nutrition from the Society for International Nutrition Research. He received a Ph.D. from Cornell University.

Laurian J. Unnevehr is a professor in the Department of Agricultural and Consumer Economics at the University of Illinois at Urbana-Champaign. Her research program is focused on the social welfare implications of food safety and diet and health linkages, including new products and new regulations. For 1993-1995 she was on leave from UIUC at the Economic Research Service of the USDA, where she worked on food safety issues. She is president-elect of the American Association for Agricultural Economics, and she received its Publication of Enduring Quality Award in 2004 for a path-breaking article on meat product demand. She is a member of the editorial board of *Food Policy* and the editorial council of the *Review of Agricultural Economics*. She received a Ph.D. from the Food Research Institute at Stanford University.

Michele Ver Ploeg (*Study Director*) was a member of the staff of the Committee on National Statistics until October 2004. In addition to the study on Enhancing the Data Infrastructure in Support of Food and Nutrition Programs, Research, and Decision Making, she directed the panel study on Estimating WIC Eligibility and Participation. Her research interests include the effects of social policies on families and children, the outcomes of children who experience poverty and changes in family composition, and individuals' education attainment choices. She received a B.A. in economics from Central College and a Ph.D. in policy analysis and management from Cornell University.

Walter Willett is professor of epidemiology and nutrition and chair of the Department of Nutrition at the Harvard School of Public Health and professor of medicine at the Harvard Medical School. He studied food science at Michigan State University, graduated from the University of Michigan Medical School, and received a Ph.D. in public health from the Harvard School of Public Health. He has focused much of his work over the last 25 years on the development of methods to study the effects of diet on the occurrence of major diseases, using both questionnaire and biochemical approaches. Starting in 1980, he applied those methods in the Nurses' Health Studies I and II and the Health Professionals' Follow-up Study. Together, these cohorts that include nearly 300,000 men and women with repeated dietary assessments are providing the most detailed U.S. information available on the long-term health consequences of food choices. His recent book for the general public, *Eat, Drink and Be Healthy: The Harvard Medical School Guide to Healthy Eating*, has appeared on major bestseller lists.